CONCEPTS OF
DIGITAL
ELECTRONICS

CONCEPTS OF
DIGITAL
ELECTRONICS

BY HARRY M. HAWKINS

TAB BOOKS Inc.
BLUE RIDGE SUMMIT, PA. 17214

To my wife Gerry, and my children Bill, Ed, Dawn, and Tom.

FIRST EDITION

SECOND PRINTING

Printed in the United States of America

Library of Congress Cataloging in Publication Data

Hawkins, Harry M.
    Concepts of digital electronics.

    Bibliography: p.
    Includes index.
    1. Digital electronics.   I. Title.
TK7868.D5H348   1983   621.381   82-19341
ISBN 0-8306-0531-2
ISBN 0-8306-1531-8 (pbk.)

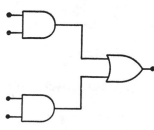

Contents

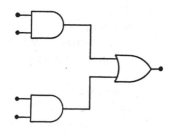

Introduction

In recent years, the application of digital electronic technology in all areas of our lives has been explosive. Industrial controls may soon run our factories. Data processing, retrieval, and storage are often handled more efficiently by digital machines than by men. Motor vehicles of all types are being equipped with digital controls. The pocket calculator has made the slide rule a museum piece. Versatile minicomputers are finding their way into homes and schools. Hand-held digital games and digital TV games are found in thousands of homes, and many home appliances are programmable (some even have memories). Nearly every area of conventional electronics has become fair game for digital technology.

Digital circuits may soon be as pervasive as vacuum tubes were thirty years ago. If we are to get maximum efficiency from these new circuits, we must have some understanding of their operation. Indeed, training in digital technology may be as important to the next generation as arithmetic is to this one, and since this new technology is changing so rapidly this learning process will need to continue throughout a person's working life.

This book offers an introduction to digital electronics. It assumes no prior knowledge of electronics, digital or otherwise. It is designed as a beginning course for the hobbyist who wishes to learn digital fundamentals.

The practical "hands-on" skills gained from the exercises in Chapters 5 through 10 will be valuable to technician and hobbyist alike. In these exercises you construct actual working models of digital functions discussed in the text. Each circuit is put through its paces and tested for proper performance. All of the exercises have been developed to provide a practical demonstration of the principle under study. The final exercise in Chapter 10 uses a number of digital principles to develop a larger system.

Each exercise has been tested and refined as a result of classroom discussion and student comments. You should fully understand the introductory material in the early chapters, which provides a brief treatment of the theory of each function, before you begin the exercises. Mathematics has been kept to a minimum and is used only where necessary to provide the basis for an electrical solution.

The hardware used in the exercises is readily available. The TTL integrated circuits, the resistors, and all other components are inexpensive and can be purchased by mail or at nearly any local electronics store. A list of suppliers is provided at the end of the book.

The following were students in my advanced digital-electronics class. Their help in the testing and revision of the exercises in this book is greatly appreciated. Without dedicated students, such as these, this book could not have been completed.

Phil Bickelhaupt
Wayne Duprez
Michael Miller
Barry O'Malley
Mark Palmesano
John Robinson

Robert Salamy
Joe Stone
Matt Thayer
Chris Wood
Ed Zak
Joe Zuber

Chapter 1

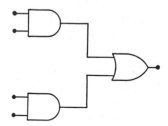

Introduction to Digital Electronics

NUMBER SYSTEMS

The number system with which we are most familiar is the base 10, or *decimal* system. It is only one of many systems, and it probably evolved because humans have a total of 10 fingers on their hands. Systems with bases other than 10 have been used by scientists for years, but their use has been limited to specialized forms of mathematics.

Recent technological developments have created the need for other number systems. The electronic computer, for example, required the development of systems that were easily adapted to electronic processes. These number systems were the *binary* (base 2), *octal* (base 8), and *hexadecimal* (base 16). The binary system is the primary language of the computer. The octal and hexadecimal systems are usually used for communication with the computer and for storage of information within the computer.

Since computers can only process binary numbers or numbers coded in other systems such as octal and hexadecimal, the decimal system must be converted to one of these other systems before it can be processed by the computer. When the computer finishes its operations on the information given to it, the output is printed or displayed in a number system other than decimal, and this too must be converted, this time back to the decimal system.

This section gives a brief explanation of the decimal, binary, octal, and hexadecimal number systems, shows how to convert

1

from one of these systems to any of the others, and gives examples of their uses in a digital system.

Decimal System

The decimal system is a number system with 10 digit possibilities, 0 through 9. Because there are 10 digits, this system is called base 10. Numbers greater than 9 are expressed as powers of 10. The number 10, for instance, is actually 10×1, and the number 100 is 10×10. Numbers expressed as powers of 10 are usually written with an exponent, or superscript. The number 10, expressed as a power of 10, is written 10^1. The number 100 is written 10^2, and the number 1000 is written 10^3. The exponent tells how many times 10 is multiplied by itself. The number 6973, for example, represents the sum of 6000 (6×10^3) plus 900 (9×10^2) plus 70 (7×10^1) plus 3 (3×10^0). Any number raised to the 0 power equals 1.

Powers of 10 is also called *scientific notation*, and it is often used to represent very large or very small numbers. The number 125,000,000,000 may be written as 1.25×10^{11}. Notice the exponent 11. In any number greater than 1, the exponent of 10 indicates how many places to the right the decimal point must be moved if the number is to be written out the long way. Numbers less than 1 are given negative exponents. The number 0.00000356 expressed as a power of 10 is 3.56×10^{-6}, or 35.6×10^{-7}. When the number is less than 1, the exponent of 10 indicates the number of places to the left the decimal point must be moved if the number is written out the long way. Table 1-1 shows the conversion of several decimal numbers to powers of 10.

In everyday transactions, it is not necessary to indicate the base of the number system we are using, because it is always base 10. In digital electronics, however, there may be more than one number system involved, and a way had to be devised for differentiating between them. The method used is the subscript method. For example, 55_{10} is 55 base 10 (or decimal), and 11_2 is 3 base 2 (or binary).

Binary System

The binary number system has a base of 2. This means that the binary system has only 2 digits, 0 and 1. In the binary number system, any number can be expressed using these 2 digits.

An example of a binary number is 110010_2. The 1's and 0's are called bits, and the position of each bit represents a power of 2. The

Table 1-1. Decimal Numbers Expressed as Powers of 10.	$1 = 1 \times 10^0$
	$10 = 1 \times 10^1$
	$100 = 1 \times 10^2$
	$1000 = 1 \times 10^3$
	$10,000 = 1 \times 10^4$
	$100,000 = 1 \times 10^5$
	$1,000,000 = 1 \times 10^6$
	$.1 = 1 \times 10^{-1}$
	$.01 = 1 \times 10^{-2}$
	$.001 = 1 \times 10^{-3}$
	$.0001 = 1 \times 10^{-4}$
	$.00001 = 1 \times 10^{-5}$
	$.000001 = 1 \times 10^{-6}$
	$654 = 6.54 \times 10^2$
	$8.50 = 850 \times 10^{-2}$
	$3.141659 = 3,141,659 \times 10^{-6}$

most significant bit (MSB) is on the left, and the least significant bit (LSB) is on the right. Figure 1-1 shows how the binary number 110010_2 is converted to 50_{10}. Notice that although the 0's in the binary number do not contribute any numerical value to the decimal equivalent, they must all appear in the binary number. It is their position that is important. For instance, if the righthand 0 is left off, the binary number becomes 11001_2. If this number is put into Fig. 1-1, the decimal equivalent becomes 25, or half of the original number.

Figure 1-2 shows how to convert from decimal to binary. The decimal-to-binary method is called the dibble-dabble method, and it involves repeatedly dividing the decimal number by 2, yielding a succession of remainders of 0 or 1. The remainders, when read in reverse order, give the binary equivalent of the decimal number. The binary-to-decimal conversion is a straightforward conversion of each binary bit to its decimal equivalent and adding these to arrive at the answer, as shown in Fig. 1-1.

The advantage of the binary number system in digital electronics is that the 2 digits, 1 and 0, can represent on or off electrical conditions. It may seem clumsy to use a number system that requires 6 bits to represent a 2 digit decimal number, but the advantages of the binary system far outweigh the disadvantages. Where there is a space problem, two other number systems, octal and hexadecimal, may be used.

Octal System

The octal number system is base 8 and comprises 8 digits, 0

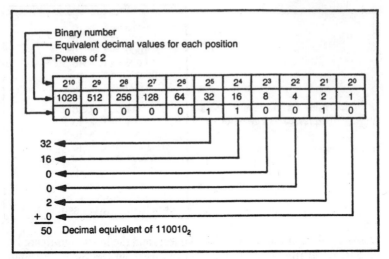

Fig. 1-1. Binary-to-decimal conversion.

through 7. A useful relationship exists between octal and digital numbers. Three binary bits count to a magnitude of 7. That is, the decimal equivalent of binary 111_2 is 7_{10}. Conversion from binary to octal, then, is simply a matter of grouping the bits of a binary number into threes beginning at the right, and then converting each group of three bits into a digit from 0 to 7. This conversion is shown in Fig. 1-3. The easiest way to convert from octal to decimal or from decimal to octal is to first convert the number to its binary equivalent. The conversion of 1432_{10} to octal is shown in Fig. 1-4.

Because octal numbers make use of 8 digits that are similar to decimal digits, they are convenient to operate with and are very

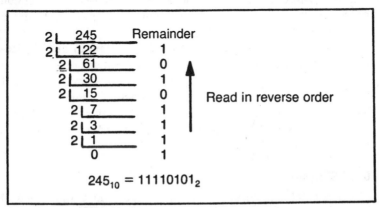

Fig. 1-2. Convert 245_{10} to binary using the dibble-dabble method.

4

Binary	101		Octal	2 7 6
Octal	5		Binary	010111110
$101_2 =$	5_8		$276_8 =$	010111110_2

Binary	110100		Octal	5 4 3 3
Octal	6 4		Binary	101100011011
$110100_2 =$	64_8		$5433_8 =$	101100011011_8

Binary	111000110001		Octal	2 0 0 1
Octal	7 0 6 1		Binary	010000000001
$111000110001_2 = 7061_8$			$2001_8 =$	010000000001_2

Fig. 1-3. Binary-to-octal and octal-to-binary conversions.

useful in computer programming. An octal number has fewer characters than its binary equivalent, and octal numbers can be converted to binary and back again by simple inspection.

Hexadecimal System

The hexadecimal system was developed as a computer programming tool. The "words" or "bytes" of information used by computers usually consist of 8, 16, 32, or more bits. Computer programs should be as short and concise as possible. Suppose that a computer programmer wants to give a computer an 8-bit instruction word. If he uses the binary system, he must type in 8 bits. If he uses the octal system, he must type in 3 digits. If he uses the hexadeci-

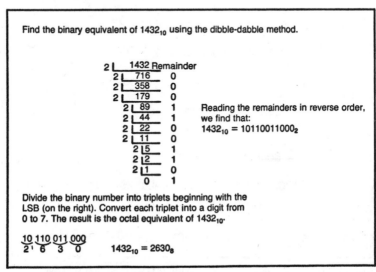

Find the binary equivalent of 1432_{10} using the dibble-dabble method.

```
2 | 1432  Remainder
  2 | 716      0
    2 | 358      0
      2 | 179      0
        2 | 89      1     Reading the remainders in reverse order,
          2 | 44      1     we find that:
            2 | 22      0     $1432_{10} = 10110011000_2$
              2 | 11      0
                2 | 5      1
                  2 | 2      1
                    2 | 1      0
                        0      1
```

Divide the binary number into triplets beginning with the LSB (on the right). Convert each triplet into a digit from 0 to 7. The result is the octal equivalent of 1432_{10}.

$$\underbrace{10}_{2} \underbrace{110}_{6} \underbrace{011}_{3} \underbrace{000}_{0} \qquad 1432_{10} = 2630_8$$

Fig. 1-4. Conversion of 1432_{10} to octal.

mal system, he only types in 2 characters. The programmer's choice is going to be the hexadecimal system because it is the shortest.

The hexadecimal system is base 16. It has 16 characters, 0 through 9 and A through F. The letters A through F are used because the numbers from 10 to 15 have two digits, and if they are used this system has no advantage over the octal system. Figure 1-5 shows the conversion of hexadecimal number $C27_{16}$ to the binary, octal, and decimal systems, and the advantages of the hexadecimal system for programming are obvious. Figure 1-6 shows the octal, binary, and hexadecimal equivalents of the decimal numbers 1_{10} through 15_{10}.

A thorough understanding of the binary, octal, and hexadecimal number systems is a must for the serious student of digital technology. The preceding was just a brief introduction. There is much more to these systems than just conversion. The information pre-

---

Each character of a hexadecimal number represents 4 binary bits.

To convert $C27_{16}$, write down the 4-bit binary equivalent of each character. (Remember, "C" is the twelfth character in the hexadecimal system.)

$$\begin{array}{ccc} C & 2 & 7 \\ 1100 & 0010 & 0111 \end{array}$$

This gives us the binary equivalent of $C27_{16}$:

$$C27_{16} = 110000100111_2$$

Divide this into triplets and convert to octal:

$$\begin{array}{cccc} 110 & 000 & 100 & 111 \\ 6 & 0 & 4 & 7 \end{array}$$

So, $C27_{16} = 6047_8$

Convert 110000100111_2 to decimal, as shown in Fig. 1-1:
$$110000100111_2 = 3111_{10}$$

Therefore, $C27_{16} = 110000100111_2 = 6047_8 = 3111_{10}$

Fig. 1-5. Convert $C27_{16}$ to its binary, octal, and decimal equivalents.

Decimal	Binary	Octal	Hexadecimal
0	0000	0	0
1	0001	1	1
2	0010	2	2
3	0011	3	3
4	0100	4	4
5	0101	5	5
6	0110	6	6
7	0111	7	7
8	1000	10	8
9	1001	11	9
10	1010	12	A
11	1011	13	B
12	1100	14	C
13	1101	15	D
14	1110	16	E
15	1111	17	F

Fig. 1-6. Four basic numbering systems.

sented in this book is sufficient for a course in digital fundamentals. More advanced courses will require knowledge of the arithmetic operations associated with these number systems. The suggested reading list at the back of this book contains a number of excellent reference texts for students who wish to gain a more thorough understanding of the binary, octal, and hexadecimal numbering systems.

CODES

Codes are systematic ways of representing information, and they are probably the oldest form of communication. Even animals often use simple codes. Human codes evolved out of a need for a way of transmitting information between people separated by physical barriers, such as distance, or by social barriers such as language. The heliograph used mirrors or lamps to flash a binary-coded message between distant points, and the telegraph used binary-coded electrical signals sent along wires. Recent developments in technology have made it necessary to develop codes for communication with machines, and the criteria for these new codes are the same as those for human-to-human codes. They must be systematic and understandable to both the machine and its operator.

The codes discussed in this unit are the codes used by machines that communicate with computers. The computer operator's job is greatly simplified if he can talk to his computer in his own language. Special machines such as the teletypewriter were developed to make this type of communication possible. These machines, called peripherals, translate the programmer's language into codes that the computer is able to use. The most common of these codes is the Binary-Coded-Decimal (BCD) code. Two other common codes are the Baudot and ASCII codes, both of which are more extensive than the BCD code, Baudot and ASCII are out-growths of the telegraph codes and lend themselves well to keyboard applications such as the teletypewriter.

Binary-Coded-Decimal (BCD) Code

BCD is a very simple code that converts decimal numbers to 4-bit binary words. Usually only the decimal numbers 0 through 9 are converted. The BCD equivalent of 5_{10}, for example, is 0101. The number 9_{10} is 1001 in BCD, and 7_{10} is 0111. Table 1-2 gives the decimal-to-BCD conversions for all the decimal numbers up through 15_{10}, which is the highest decimal number that can be represented by a 4-bit digital word.

Calculators are an example of the BCD code in use. Each of the number keys, when depressed, sends a BCD signal for the number that the key represents to the internal digital circuitry of the calculator. Key number 4, for instance, will send 0100 into the calculator. When the desired calculation is completed, the calculator output is converted from BCD back to decimal for display.

Decimal	BCD
0	0 0 0 0
1	0 0 0 1
2	0 0 1 0
3	0 0 1 1
4	0 1 0 0
5	0 1 0 1
6	0 1 1 0
7	0 1 1 1
8	1 0 0 0
9	1 0 0 1
10	1 0 1 0
11	1 0 1 1
12	1 1 0 0
13	1 1 0 1
14	1 1 1 0
15	1 1 1 1

Table 1-2. Decimal
Converted to BCD Code Equivalents.

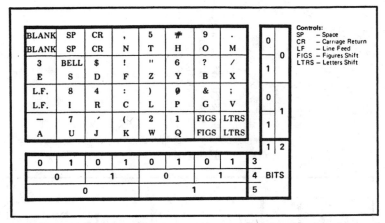

Fig. 1-7. Baudot code. Five-level code bits are obtained from bottom lines and right side. For example, C = 01110 (courtesy the Teletype Corp.)

Baudot Code

The Baudot code (pronounced bah-doe) was developed for use with teleprinter machines. The BCD code will not work with these machines because it can represent only sixteen characters, and these machines normally have many more than sixteen keys. The Baudot code was a natural extension of the telegraph codes already in use. Figure 1-7 shows a typical teleprinter keyboard and the Baudot code for each key. With this code, 5 digital bits can represent 32 different keyboard characteristics.

Information is typed into the keyboard of the teleprinter. The machine automatically codes each keystroke into a series of 5 binary bits. A receiving machine converts the stream of Baudot-coded data back into typed copy. Each 5-bit word of data has pulses added to it that separate one word from another. In Fig. 1-7 these pulses are represented by the space between the keyboard drawing and the code bits to the right and below it. Each word, then, begins with a space, and the next 5 bits are the Baudot code that represents the character of the key that has been depressed. A stop pulse that is somewhat longer than the start pulse is added to the end of the word. The Baudot code is called a 5-level code because each character of transmitted data contains 5 binary bits.

ASCII Code

ASCII (pronounced as-kee) is the acronym for the American Standard Code for Information Exchange. This code is the successor to the Baudot code and has several important advantages over it.

Fig. 1-8. ASCII code showing 7-bit (level) make-up. Bits are arranged along the bottom and right side. For example, D = 0010001 (courtesy the Teletype Corp.)

Controls		Characters					
NUL	DLE	SP	0	@	P	`	p
SOH	DC1	!	1	A	Q	a	q
STX	DC2	"	2	B	R	b	r
ETX	DC3	#	3	C	S	c	s
EOT	DC4	$	4	D	T	d	t
ENQ	NAK	%	5	E	U	e	u
ACK	SYN	&	6	F	V	f	v
BEL	ETB	'	7	G	W	g	w
BS	CAN	(	8	H	X	h	x
HT	EM	)	9	I	Y	i	y
NL	SUB	*	:	J	Z	j	z
VT	ESC	+	;	K	[	k	{
FF	FS	,	<	L	\	l	¦
CR	GS	-	=	M	]	m	}
SO	RS	.	>	N	^	n	~
SI	US	/	?	O	_	o	DEL

BITS

Controls:
NUL – Null
SOH – Start of Heading
STX – Start of Text
ETX – End of Text
EOT – End of Transmission
ENQ – Enquiry
ACK – Acknowledge
BEL – Bell
BS – Back Space
HT – Horizontal Tabulation
NL – New Line
VT – Vertical Tabulation
FF – Form Feed
CR – Carriage Return
SO – Shift-Out
SI – Shift-In
DLE – Data Link Escape
DC1 – Device Control 1
DC2 – Device Control 2
DC3 – Device Control 3
DC4 – Device Control 4
NAK – Negative Acknowledge
SYN – Synchronous Idle
ETB – End of Transmission Block
CAN – Cancel
EM – End of Medium
SUB – Substitute
ESC – Escape
FS – File Separator
GS – Group Separator
RS – Record Separator
US – Unit Separator

The Baudot code could only represent 32 functions or characters, and these were not in ascending binary order. These two limitations of the Baudot code made some computer operations very difficult, and others impossible. The ASCII system largely eliminates these problems.

Figure 1-8 shows an ASCII keyboard and the complete ASCII code. This code, like the Baudot code, is particularly well suited for teleprinter keyboards, and the ASCII code is extensive enough to make it an excellent keyboard-to-computer code.

In Fig. 1-8, the right-side 4 bits and the three bits along the bottom are used to develop each keyboard character or function. Since the ASCII code uses 7 bits to represent each character, it is known as a 7-level code. An eighth bit is sometimes added as a "parity" bit. This parity bit is either added or left out so that the complete character will have either an even or odd number of "on" or "mark" positions. A parity check is a way of detecting errors in a transmitted message. If the computer receiving an ASCII-coded message has been told to accept only words with an even number of marks, it will reject any with an odd number and notify the operator that there has been a transmission error.

Figure 1-9 shows a typical ASCII word. Notice that a start pulse is provided at the beginning, and a double mark pulse is included as a stop pulse. The 7 pulses in between carry the actual ASCII code, which in this example is the keyboard character "U." The parity bit, if used, would occupy the eighth position, and it would be an "on" or mark, pulse.

Most modern computers use the ASCII code for communication with the outside world. Using this code, computer operators can type messages in their own alphabet and the keyboard circuitry

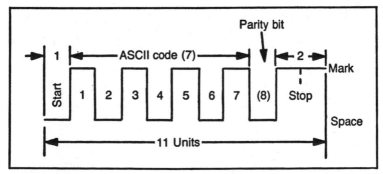

Fig. 1-9. Typical 7-level ASCII code in 11-unit form with parity bit used for teleprinters.

translates these into ASCII for the computer. Transmissions from the computer are translated from ASCII back into the operator's language.

There are other codes, such as the Extended Binary-Coded Decimal Interchange Code (EBCDIC), used mostly by large computers, and the Hollerith code (the code used to store data on punched cards), but these codes are not generally used in elementary digital operations such as those discussed in this book.

Chapter 2

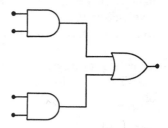

Basic Digital Operations

This section deals with basic digital operations and the special circuits used to implement these operations. Digital electronics is a "logical" science. Logic, generally speaking, is the science of formal principles of reasoning. *Digital* logic is the science of reasoning with numbers; it is an "if, then" science in the most literal sense of the words. *If* a particular set of circumstances occurs, *then* a specific action results. The result is always the same for a given set of circumstances.

This predictability in digital logic simplfies digital electronics considerably. Nearly all digital functions can be performed by a special circuit called a *gate*. If the logic operation is too complex for one gate, it can almost always be implemented through the use of a combination of gates. These extended logic circuits are called *combinational logic.*

This chapter explains the basic logic functions of AND, NAND, OR, NOR, and exclusive OR. Combinational logic and the inverter and flip-flop functions are also covered. After reading this section, you should begin performing the exercises found in Chapter 5. They will provide a step-by-step verification of the information given in this section.

"AND" AND "NAND" FUNCTIONS

Two of the most basic logic functions are the AND and NAND functions. The difference between these functions is that they are complements. This means they are opposite in function.

The electronic manipulators of digital functions are called *gates,* and these are used to control digital data in a number of ways. The word gate is applied to these devices because they are often used to stop and start a stream of digital data—much like opening or closing a "gate."

Positive logic will be used in all illustrations in this book to help clarify the operating principles. Positive logic means that a high level, a one (1), or a positive voltage, is true (or on). A low level, a zero (0), or an absence of a positive voltage, is false (or off). (Negative logic is also used in digital electronics. This is sometimes interpreted as being the opposite of positive logic, i.e., positive (+) is low and negative (−) is high. A more accurate description is to say that, in negative logic, a low (0) is true or on, and a high (1) is false or off.)

AND Gate

The AND gate provides a function in digital logic which gives a high (+) output when all of its inputs are high (+). Figure 2-1 shows the symbol used to represent an AND gate. In this case, there are two inputs, A and B, and one output, C. Gates with as many as eight inputs are available.

Figure 2-2 shows how the AND gate works. This is a simple circuit consisting of lamp (C), two switches (A and B), and a battery. The switches are the inputs and the lamp is the output. The principle is that a certain output condition will always occur when the inputs are placed in certain positions. Each input has only two conditions—on or off. On is considered a 1 (or high), and off is considered a 0 (or low).

The truth table in Fig. 2-3 shows all the possible input conditions or logic states, and the resulting output or logic state for each. (The truth table (Fig. 2-3) shows the logic of each state to be as follows:

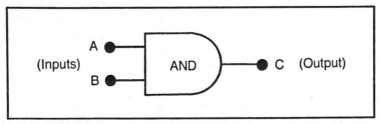

Fig. 2-1. Logic symbol for a 2-input AND gate. The AND is not usually written in the symbol.

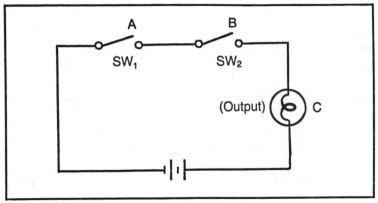

Fig. 2-2. Electrical equivalent of a logic AND gate.

State 1. Both A and B are "open" or in logic state 0. No current can flow so the lamp (C) output is off or in a zero (0) or low state.

State 2-3. In both state 2 and 3, one switch is closed (1), and the other remains open (0). This is an incomplete circuit and lamp C (output) will be in a low (off or 0) condition in both of these states.

State 4. In this state *both* switches are "on" or in a logic high state. The circuit is complete and lamp C (output) is in an "on" or high (1) logic state.

It can be seen from this procedure that an AND gate will provide an output (1) *only* when *both* inputs (or all inputs if there are more than two) are high (+). This relationship is usually written in the following way:

$$A \cdot B = C \text{ or } AB = C$$

State	Input		Output
	A	B	C
1	0	0	0
2	1	0	0
3	0	1	0
4	1	1	1

Fig. 2-3. AND-gate truth table showing all of the possible input logic conditions and the resulting output logic states.

The dot between A and B indicates an AND function. This kind of relationship is a mathematical formula that falls under the general heading of Boolean algebra. This is a special type of mathematics that is used a great deal in digital logic. The references listed at the end of this book all contain excellent information on Boolean algebra and should be consulted if more information is desired. It is not necessary to understand Boolean algebra in order to understand the basics of digital electronics. The need for this type of algebra increases as one becomes more skillful in the area of digital electronics.

The expression AB = C is normally said in the following way:

"If A equals 1 and B equals 1, then C will equal 1"

or

"A and B equal C"

As a practical example of an AND gate application, suppose an AND gate were attached to sensors as shown in Fig. 2-4. A temperature sensor provides switching between 69° and 70° F so that if the temperature is 70° F or higher, a high (1) will be applied to input A. Input B will receive a high (1) if the rain sensor is dry (no rain). The following statement may be used to show how this logic circuit can be used to provide a signal at C (light on or off) that can aid in making

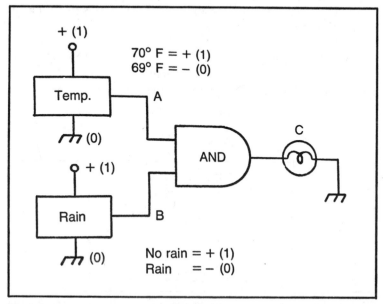

Fig. 2-4. Logic diagram showing AND decision made with temperature and moisture input sensors.

Input		Output	
A	B	C (AND)	C(NAND)
0	0	0	1
0	1	0	1
1	0	0	1
1	1	1	0

Fig. 2-5. Truth table showing both AND and NAND logic functions.

a decision: If the temperature is 70° F or more, AND it is not raining, then the light is on and the weather is good for swimming. If the lamp is off, conditions are not good for swimming and some other activity should be selected.

The point here is that any logic decision that requires a series of conditions to be present in order to cause a decision to be made can be done electrically using AND gates. Of course, the example used here can be done more easily without any formal gates, but it does demonstrate the principle. In practice, transistor-type switches are used rather than the manual toggle switches shown (Fig. 2-2).

NAND Gate

The NAND gate is a more common gate than the AND gate. It is the complement of the AND gate, meaning that it is just the opposite in logic state.

NAND gates frequently are cheaper and easier to use than AND gates because electronically they are simpler. Later in this chapter AND gates will be constructed using NAND gates.

In Fig. 2-5, the truth tables for both AND and NAND gates are shown. The N in front of the AND means "NOT" AND.

The logic is also written differently for a NAND gate when it is placed in a formula. An overscore or bar is used to indicate a NOT or inverted (complement) condition. For example,

$$A \cdot B = \overline{C}$$

is the expression for the NAND function in Fig. 2-5. It is spoken as: A and B equals NOT C.

In the case of the NAND gate, the output will be high (1) *except* when the inputs are all *high*. When all inputs are high, the output of the NAND gate will be low (0).

The example of a decision about going swimming could be used with a NAND gate as well. In this case, the light would need to be "off" in order to indicate that conditions are suitable for swimming. If the lamp is on, it would be a signal that one of the conditions is not suitable and another activity must be selected.

The symbol used for the NAND gate is nearly the same as the AND gate symbol. The difference is in the addition of a small circle at the point where the output leaves the symbol, as shown in Fig. 2-6. The circle means that whatever the expected output of the AND gate is, it will now be inverted, or just the opposite of the normal logic state. This symbol (the small circle) is used on other gates and logic device symbols as well. It can be used on inputs to indicate that the logic state of the input must be inverted (or will be inverted in the device).

As mentioned earlier, the NAND gate is the most common of all the gates. It can be used to create other logic functions such as the AND gate, and for that reason it is a valuable device. More about the use of the NAND gate will be presented later in this chapter.

OR AND NOR FUNCTIONS

The OR gate is as important to the fucntion of digital logic circuits as the AND gate. The OR and NOR functions are complements of each other. These gates provide the logic function that represents conditions when any one of a number of events occurs, an action will result. As with other gates, these functions are used to steer or direct, logic.

OR Gate

The OR gate provides a function that will give a high output (1) when any one of the inputs is at a high (1) logic level. Figure 2-7 shows the symbol commonly used to represent the OR gate. In this case, only two inputs are shown. Gates often have more than two inputs. A and B represent the two inputs and C is the output.

The OR function is electrically shown in Fig. 2-8. The two switches represent inputs A and B and lamp C represents the

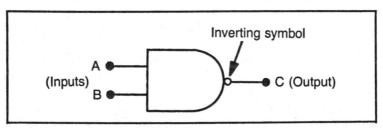

Fig. 2-6. Symbol for the NAND gate.

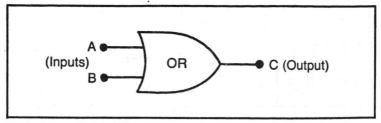

Fig. 2-7. Logic symbol for a 2-input OR gate.

output. Figure 2-9 shows all the possible logic conditions of both the inputs and output. From this truth table, it can be seen that a high output (logic 1) appears at C if either of the input switches is in the closed (logic 1) position. Closing either of the switches will light lamp C. The truth table shows the following:

State 1. Both A and B are low (logic 0), so the circuit is incomplete and lamp C is off (C = 0).

States 2-3. In both of these states one switch is on and the other off. In each case, the lamp C is ON due to the logic 1 at C.

State 4. Both switches A and B are on, producing a logic 1 at C and lighting the lamp.

It can be seen from this analysis that the output C will be at a logic 1 condition at anytime one OR the other of the inputs is at a

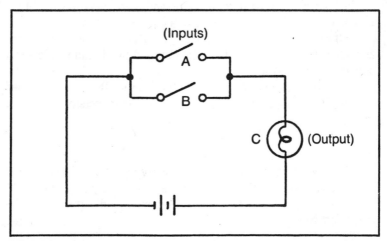

Fig. 2-8. Circuit showing the logic OR function.

State	Inputs		Output
	A	B	C
1	0	0	0
2	1	0	1
3	0	1	1
4	1	1	1

Fig. 2-9. Truth table for logic OR function.

logic 1 state. This relationship is usually written in the following way:

$$A + B = C$$

The plus (+) sign is the symbol used to indicate an OR logic function. This expression is spoken in the following way: "If A OR B equals 1 then C equals 1."

As an example of an application of the OR function, consider a case where either of two bits of positive information is needed to make a positive decision. Suppose you wish to attend a basketball game but only if one or both of your friends will also attend. If neither friend can attend, then you will not attend. Suppose a circuit is wired between the three houses as shown in Fig. 2-10. A switch is located at the house of each of the two friends and the lamp (decision) is located at your house (C). At a pre-determined time, say 5:00 PM, each friend will decide whether or not to go to the game. Each places the decision switch (A and B) in the yes (on) or no (off) position. The decision lamp (at C) is activated (power is applied to the OR gate) and the result is determined. If lamp C is not lit, then neither friend wants to go to the game and therefore you will not go. If lamp C is lit, one or both friends will go and therefore so will you.

Of course, there is no way to tell if only one friend wants to go, and which one it is, or if both will be going. A phone call would be far superior to this system.

The important point in all of this is that the OR gate can be used to steer logic states and to communicate data for decision making or informational purposes.

NOR Gate

The NOR gate is another commonly used logic circuit. This gate simply produces the *inverted* state of the OR gate. Figure 2-11 shows the logic state conditions of inputs and outputs for the two-input NOR gate.

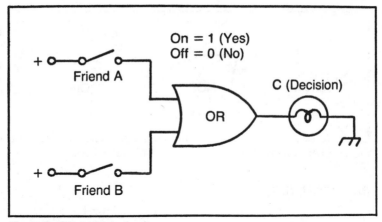

Fig. 2-10. OR decision circuit.

The N in front of the OR means NOT OR, or just the opposite state of logic as is supplied by the OR gate. Notice that this relationship is similar to the relationship between the NAND and AND gates. Figure 2-12 shows the symbol used for the two-input NOR gate. The small circle at the output changes the OR symbol into the NOR symbol, and the correct way to express this NOR relationship is:

$$A + B = \overline{C}$$

This is spoken as: A OR B equals NOT C. It may also be said in the following way: If A is high (1) OR B is high (1) then C (output) must be NOT HIGH (0). The use of the overscore or bar over the letter indicates an inverted logic state.

Although gates are used for specific decision-making purposes and can stand alone for this purpose, they can be used for complex logic decisions as well. When gates are combined in various ways, very complex logic functions can be the result.

Fig. 2-11. Truth table showing logic for 2-input NOR gate.

State	Inputs		Output
	A	B	C
1	0	0	1
2	0	1	0
3	1	0	0
4	1	1	0

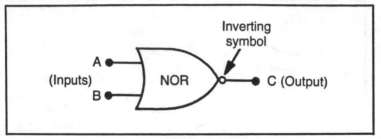

Fig. 2-12. Logic symbol for a 2-input NOR gate. The NOR is not usually written on the symbol.

INVERTER FUNCTIONS

The process of inverting, or simply changing a 1 to a 0, or a high logic state to a low logic state, is an important function in digital electronics. Since a binary number system is usually in use, the inverter is used to change from one logic state to its complement, or opposite, logic state. This unit shows how this process is accomplished using gates and other devices as inverters.

Gate Inverters

All digital integrated circuit (IC) gates such as the NAND and NOR can be used to invert a logic state. Consider the NAND gate in Fig. 2-13. Notice that both inputs are tied together so that whatever logic is applied to the input (X) will be applied to both A and B at the same time. The truth table in Fig. 2-14 shows the two possible combinations that can occur in this case. Notice that if both A and B are low (0) then the output (C) is high (1), and that when both A and B are high (1) then the output (C) is low (0). This shows that by applying the same logic condition to *all* of the inputs of a NAND gate (or a NOR gate), the output will be inverted. If the NAND or NOR gate has many inputs, all must be tied together and treated as a single input.

Figure 2-15 shows the logic symbol used for an inverter. The small circle at the output indicates the inverting function. The

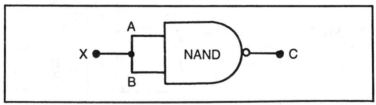

Fig. 2-13. NAND gate inverter.

X		C
A	B	Output
0	0	1
1	1	0

Fig. 2-14. Truth table showing two possible input and output states for NAND gate inverter.

triangle shape is the standard electronic symbol for an amplifier or buffer.

In some cases where a non-inverting *buffer* is needed, an AND or OR gate may be used. A buffer may amplify but it usually simply provides a separation between two circuits so that one circuit does not interfere with the other. Figure 2-16 shows the symbol for a buffer (non-inverting). Notice that the inverting circle is not present at the output.

Figure 2-17 shows the buffer logic. In both states, the input(s) are either high (1) or low (0) since all inputs are common. Under these conditions, a low (0) input (A and B) will produce a low (0) output (C) and a high input (A and B) will produce a high (1) output at C. This shows that if an AND gate is used, it will not invert but it will serve as a buffer. This function is also true for the OR gate.

Other Inverters

Inverters are frequently needed when processing digital information. Integrated circuits that contain six inverters (the 7404 for example) are available at low cost. In cases where some NOR or NAND gates are "left over" or not used in a circuit, the unused gates can be used as inverters. It makes sense to use a NAND or NOR gate that is not used for any other purpose rather than install a new IC in order to obtain a single gate or inverter.

In some cases, where only one inverter is needed, it may be cheaper and easier to use a single transistor inverter. A later chapter in this book provides more detail on the switching and inverting applications of the transistor.

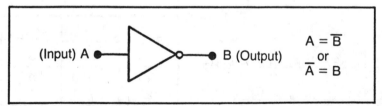

$$A = \overline{B}$$
or
$$\overline{A} = B$$

Fig. 2-15. Inverter logic symbol and logic relationship.

23

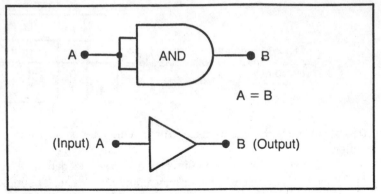

Fig. 2-16. Symbol for non-inverting buffer or amplifier.

COMBINING LOGIC FUNCTIONS

Individual logic gates are building blocks. They can stand alone if only a single logic function is needed, or they can be combined with other gates for more complex operations. There are times when it is advantageous to substitute one kind of gate for another. What does a circuit designer do, for instance, if a NAND gate is needed and only AND gates are available? It is an indicator of the versatility and flexibility of digital integrated circuits that the designer can configure the gates on hand to meet the circuit needs. This unit discusses logic function substitution and combinational logic, and shows how to configure one kind of circuit so that it will do the job of another.

NAND-AND

A quick look back at the truth table for the NAND gate (Fig. 2-5) will show that by inverting the output of a NAND gate we get the AND function. So the designer who needs an AND and has only NANDs need only invert the outputs. Figure 2-18 shows the diagram of this operation, and Fig. 2-19 is the truth table. Notice that the "D" column of Fig. 2-19 is exactly the same as the output of the AND function in Fig. 2-3.

Fig. 2-17. Truth table for buffer logic.

| State | Input | | Output |
	A	B	C
1	0	0	0
2	1	1	1

24

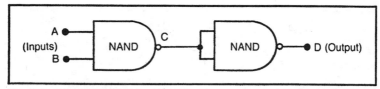

Fig. 2-18. AND function using NAND gates.

Figure 2-20 shows the way this circuit can be redrawn. All three illustrations are equivalent, since each performs the same logic function (AND), so any one is correct. The simple form of the AND gate symbol (Fig. 2-20C) will serve most purposes. The use of the complete gate diagram as shown in Fig. 2-20A indicates the way in which the circuit is wired and may be more useful for that reason. It is important to remember that, by using an inverter at the output, AND gates can be made from NAND gates. The process works as well when using an AND gate to make a NAND, but is less commonly used.

NOR-OR

The NOR gate can be used to develop an OR function in the same way. The output of the NOR gate or OR gate, when inverted, provides the logic function of its complement.

Figure 2-21 shows this principle. Notice that all three illustrations (a, b, and c) are equal in logic function. Figure 2-22 shows the truth table for this function. Notice that the NOR and OR functions are simple inversions (complements) of each other.

In practice, Fig. 2-21C is the only symbol necessary to describe this logic function. However, because of the use of NOR gates, Fig. 2-21A, may be more suitable since it also shows how the function is obtained. All three diagrams are technically correct and any one can be used to express the OR function.

If an OR gate is used, and its output is inverted (by using a NAND or NOR gate or inverter), then the result is NOR gate. In

Fig. 2-19. The 2-input NAND and AND gate truth table comparison.

Input		NAND	AND
A	B	Output C	Output D
0	0	1	0
0	1	1	0
1	0	1	0
1	1	0	1

25

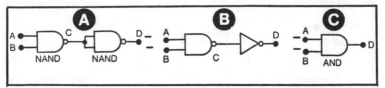

Fig. 2-20. AND function using NAND gates.

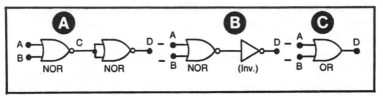

Fig. 2-21. OR function using NOR gates.

practice, the NOR gate is more often used to develop the OR function than the other way around. This is because the NOR gate is more common and usually costs less.

NAND-OR

So far, the *outputs* of gates have been inverted in order to obtain desired logic functions. The *inputs* of NAND and NOR gates can also be inverted in order to obtain desired logic functions.

In Fig. 2-23, two NAND gates are used as inverters to invert the inputs to a third NAND gate. The result, according to the truth table, Fig. 2-24, is the OR gate function. The OR function is shown in the right side column of Fig. 2-24 so that it can be compared with the output C; it is identical to C.

The importance of this function is that if the inputs to a NAND gate are inverted, the result is an OR gate function. This is a useful bit of information since it provides another way (other than purchasing an OR gate IC) to develop the OR function with the use of the common NAND gate IC. The inverting function is shown sym-

Input		NOR Output	OR Output
A	B	C	D
0	0	1	0
1	0	0	1
0	1	0	1
1	1	0	1

Fig. 2-22. Truth table showing relationship between logic of NOR gates used to develop an OR function.

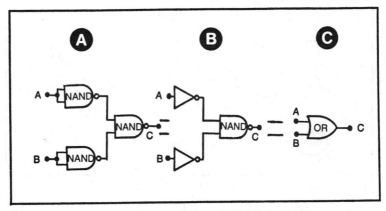

Fig. 2-23. OR function using NAND gates.

bolically in Fig. 2-23. This inversion function doesn't necessarily have to be performed with a NAND gate; a NOR gate inverter or any other inverter may be used. Figure 2-23A is the preferred drawing since it shows the way in which the OR function was obtained. Figure 2-23C, however, is really all that is necessary to show the logic function that is being used.

NOR-AND

If the inputs to a NOR gate are inverted, the logic result is an AND gate function. Figure 2-25 shows the progression of this circuit with NOR gates. Again, all three (A, B, and C) illustrations are equivalent since each provides the same logic function (AND). Figure 2-25C is all that is needed to show the logic, but Fig. 2-25A is more descriptive of how the function is obtained. Any inverter can be used to invert the input logic states.

Figure 2-26 shows the truth table of this gate system. The AND gate truth table is included in the right-hand column. Notice that the output C is exactly the same as the AND gate logic.

Fig. 2-24. OR function truth table resulting from inverting the inputs of a NAND gate.

Inputs		Output C	OR Function
A	B		
0	0	0	0
0	1	1	1
1	0	1	1
1	1	1	1

27

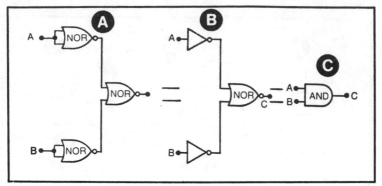

Fig. 2-25. AND function using NOR gates.

COMPLEX LOGIC FUNCTIONS

It is frequently necessary to electrically carry out logic functions that contain a combination of many gates of different types. Decisions that require a complex input can be made using either sequential or combinational logic. The sequential-type logic involves a series of events or decisions that must proceed in a one-following-another, or serial fashion. Combinational logic generally involves combining a number of logic functions so that the final output or decision will be based on the functions performed by all the combinations. The design of complex logic functions is usually a result of applying the principles of Boolean algebra to the problem. This form of mathematics is specifically geared to deal with the binary number system, and it is useful in reducing logic circuits to a minimum number or other operations, thereby reducing the cost of a circuit.

Combinational Logic

Combinational logic refers to a type of electrical circuit that is made up of various gates. The final output, or decision, that results from this circuit depends on the type of gates used, how they are combined, and the inputs that are applied. Frequently, com-

Input		Output C	AND Gate Logic
A	B		
0	0	0	0
0	1	0	0
1	0	0	0
1	1	1	1

Fig. 2-26. Truth table showing AND function derived from NOR gates.

binational-logic circuits that have a wide use are packaged in single IC units. Integrated circuits such as encoders, decoders, and display drivers are examples of combinational logic ICs.

An example of a combinational logic circuit is shown in Fig. 2-27. In this case, there are four inputs; A, B, C, and D. The final output, E, represents a final result or decision. Figure 2-28 shows the logic or truth table for this circuit. The truth table provides all 16 possible combinations that can occur with the four inputs. Notice that only at steps 4, 8, 12, 13, 14, 15 and 16 is there a high (1) output at E. This circuit then provides an output (1) only when A AND B OR C AND D are in a high (1) logic condition, and at no other time. The Boolean formula for this circuit is:

$$(AB) + (CD) = E$$

This example can be related to an actual decision-making event such as the following:

> Suppose an individual is planning a date that requires the use of an automobile, and the date is to be with a second couple. If couple X (input A AND B) or couple Y (input C AND D) agree to go, then the date is on. Both people in one or the other or

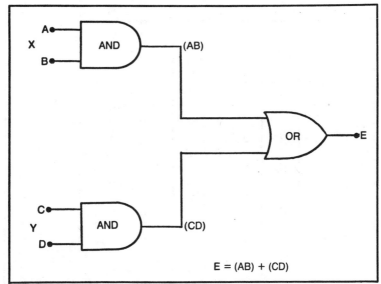

Fig. 2-27. Combinational logic circuit using two 2-input AND gates and one 2-input OR gate.

both of the couples must go. Each couple has access to a car so that either couple can go even if the other couple does not go. The decision making for this double date is relatively simple and it can be duplicated with the circuit shown in Fig. 2-27 and the truth table, Fig. 2-28.

This example may seem to be a complicated solution to a simple problem, but it is used only to show the logic used in the process. Combinational logic decisions of this sort, but much more complex, are made by individuals every day without formalizing the process with a truth table. As problems and decisions become more complex, the use of a truth table (and perhaps an automatic electrical gate system) becomes more useful.

Sequential Logic

Sequential logic circuits are usually used where a decision must be made based on previous information and new or updated information. Usually, a sequence of events must occur, often in a

Step	Input				Output
	A	B	C	D	E
1	0	0	0	0	0
2	0	0	0	1	0
3	0	0	1	0	0
4	0	0	1	1	1
5	0	1	0	0	0
6	0	1	0	1	0
7	0	1	1	0	0
8	0	1	1	1	1
9	1	0	0	0	0
10	1	0	0	1	0
11	1	0	1	0	0
12	1	0	1	1	1
13	1	1	0	0	1
14	1	1	0	1	1
15	1	1	1	0	1
16	1	1	1	1	1

Fig. 2-28. Truth table for the logic diagram in Fig. 2-27, a combination of two AND gates and one OR gate.

Step	Inputs		OR Gate	EXCLUSIVE OR Gate
	A	B		
1	0	0	0	0
2	0	1	1	1
3	1	0	1	1
4	1	1	1	0

Fig. 2-29. Truth table for 2-input OR gate and exclusive-OR gate.

definite order, before an output can occur. A timing and memory function is usually employed with this kind of logic.

Counters, shift registers, and memory devices are used in sequential logic. In some cases, a stored memory is added to (updated) by inputs that occur on a regular (clocked) basis. The output may then be a function of both the stored and the newly received data.

Most sequential-logic circuits are called counters and shift registers. These units are available in a wide variety of types in digital integrated circuits. In most cases, clocks or oscillators (multivibrators) are a part of the sequencial circuit. The process is usually controlled or "clocked" by a multivibrator or square-wave generator. More will be said about these circuits later.

EXCLUSIVE-OR GATE FUNCTION

A special gate known as the exclusive-OR (XOR) gate has some special applications in digital logic circuits. This gate is used in digital adding or arithmetic circuits. It is also used in decoder circuits, especially in error detection and digital-word identification.

Exclusive-OR

The standard OR gate provides a high output (logic 1) when any one or more of the inputs is at a logic 1 level. The exclusive-OR gate, however, will provide an output (logic 1) only when the two inputs are *not alike*. This means that the two inputs must be different in logic level in order for logic 1 output to occur. Figure 2-29 shows the standard OR and the exclusive-OR logic truth table. Notice that, for the exclusive-OR gate, and output (logic 1) results *only* in steps 2 and 3 where the inputs (A and B) are different in logic state.

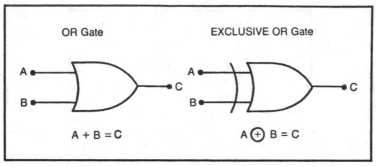

Fig. 2-30. Symbols of a standard OR gate and an exclusive-OR gate.

The logic symbol for the exclusive-OR gate is shown in Fig. 2-30. Notice that a double bar is placed at the input side. The standard OR gate symbol is shown for comparison.

The Boolean expression for the exclusive OR gate uses the plus (+) sign as does the regular OR gate symbol, but it has a circle drawn around the plus, as shown in Fig. 2-30.

Parity Checking

One important use of the exclusive-OR gate is in checking for parity. Parity, in digital electronics, means the number of high (logic 1) states that can occur in a digital word. Parity can be either odd or even. For example, in Fig. 2-31, two digital words are shown. The word in Fig. 2-31A has an even number of 1s (high states) and is an *even-parity* word. The other example has an *odd* number of 1s and is an *odd-parity* word.

The circuit shown in Fig. 2-32 is a parity checker that is used to identify an odd parity in a four-bit binary word. The truth table in Fig. 2-33 shows the logic states under all input conditions for this circuit. A parity column is used to indicate those words (8 in all, or half of the total) which have an odd parity (odd total number of 1s). It can be seen from Fig. 2-33 that anytime a four-bit binary word with an odd parity is applied to the inputs of the circuit in Fig. 2-32, a high

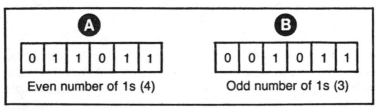

Fig. 2-31. Two 6-bit digital words showing both odd and even parity.

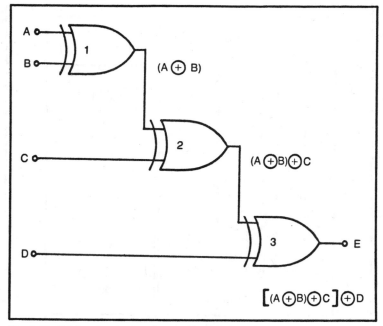

Fig. 2-32. Odd-parity checker for a 4-bit digital word.

output from the checker will result. To be practical, suppose the data being sent uses four-bit binary words but the words are selected to always have an *even* parity. As each word is presented to the parity checker, no output will occur if the word has an *even* parity. This means all is well. If, however, interference of some sort changes a word (by adding or taking away a pulse or two) the result may be a word with an *odd* parity. If this happens, it represents an error. The parity checker can be used to immediately stop the transmission process and signal an error, or it can be used to print an error message. It can also be used to print an error symbol in place of the character that would normally appear.

More elaborate parity checkers can be constructed for a variety of other uses, such as correcting tests. Any answer that does not meet the parity requirement previously set is an error.

Word Detector

Another application of the exclusive OR gate is in binary "word" detection. Figure 2-34 shows a four-bit word detector using two exclusive OR gates and one OR (or AND) gate. The truth table for this type of circuit is given in Fig. 2-35. Examination of this table

| Inputs | | | | Output | Parity |
A	B	C	D	E	(number of 1's)
0	0	0	0	0	
0	0	0	1	1	odd (1)
0	0	1	0	1	odd (1)
0	0	1	1	0	
0	1	0	0	1	odd (1)
0	1	0	1	0	
0	1	1	0	0	
0	1	1	1	1	odd (3)
1	0	0	0	1	odd (1)
1	0	0	1	0	
1	0	1	0	0	
1	0	1	1	1	odd (3)
1	1	0	0	0	
1	1	0	1	1	odd (3)
1	1	1	0	1	odd (3)
1	1	1	1	0	

Fig. 2-33. Truth table for odd parity checker for a 4-bit digital word. Only the odd parity words are indicated.

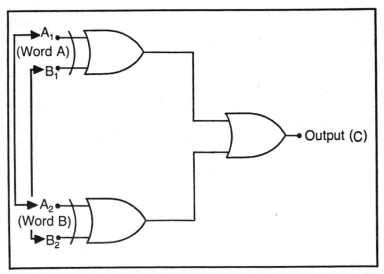

Fig. 2-34. A word comparator for two 2-bit binary words using exclusive-OR gates.

34

Steps	Word A		Word B		Output C
	A_1	A_2	B_1	B_2	
1	0	0	0	0	0
2	0	1	0	1	0
3	1	0	1	0	0
4	1	1	1	1	0
5	0	0	1	1	1
6	1	0	0	1	1
7	0	1	1	0	1

Fig. 2-35. Truth table for 2-bit binary word detector.

shows that when words A and B are identical (steps 1 through 4) the output at C is low. If, however, the words are *not* identical, as in steps 5, 6, and 7, the output at C is high.

This type of word comparison is very helpful when searching through a large amount of data (words) for a particular word. The number of times the word occurs can be counted with this circuit, providing a frequency count of the word. For example, suppose the problem is to find out how many people in a given population (data base) are 5'7" tall. If the data base has all heights coded into digital words, and the word for 5'7" is 00110, the word detector will supply a high (1) output each time this particular digital word is encountered as the data base is passed through the detector. The output can be counted and displayed. There are, of course, many other applications, such as digital combination locks, that employ the exclusive-OR gate principle.

Adding

One of the most basic functions of computing equipment is the addition of binary numbers. In fact, every arithmetic operation carried out by a computer is done by addition, including multiplication and division. These last two are accomplished by a complex system of add operations and what are called "shifts." The exclusive-OR gate is usually the basic unit of computer adders. This is because these gates are good *comparators;* they compare two or more electronic signals and decide whether they are alike or different. This was the purpose of the exclusive-OR gate in the word detector, and it is also its use in binary adders. Binary addition is not true addition; since there are only two digit possibilities, it is merely necessary to compare inputs to an adder to arrive at a sum. This comparison is what the exclusive-OR gate does best.

Binary addition is not difficult to understand if we remember that the binary number 10_2 is the same *type* of number in the binary system as 10_{10} is in the decimal system. They are not *equivalent;* if 10_2 is converted to decimal, the result is 2_{10}. What they do have in common is that they are both the first double-digit number in their respective systems. Using this as a rule-of-thumb, the rules for binary addition are the same as those for decimal addition. In decimal addition, any time the sum of two numbers is greater than 9, the least-significant digit is brought down as a sum and the most-significant digit is carried over to the next column. The principle is the same in binary addition. Any time the sum of two numbers is greater than 1, the least-significant digit (0) is brought down as a

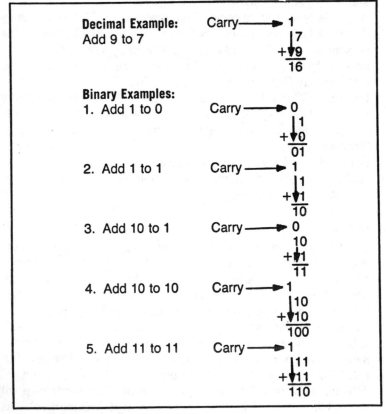

Fig. 2-36. Examples of binary addition.

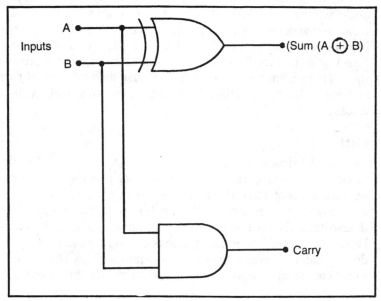

Fig. 2-37. Half-adder circuit using exclusive-OR gate and AND gate.

sum, and the most- significant digit is carried to the next column. Several examples of binary addition are given in Fig. 2-36.

Figure 2-37 shows the logic diagram of a 2-bit binary adder. This is the simplest form of adder, since it can only add two binary bits together, and it is called a *half-adder* because it has no input to receive the carry from a previous adder. A *full-adder* is one with a carry input. Figure 2-38 gives the truth table for the half-adder shown in Fig. 2-37. Notice that the only time there is a carry is in step 4, when both inputs are 1s.

CLOCKS AND MULTIVIBRATORS

Proper and accurate timing is absolutely essential in digital circuits. In a pocket calculator, for example, a relatively simple

Fig. 2-38. Truth table for half-adder circuit.

Step	Inputs		Outputs	
	A	B	Sum	Carry
1	0	0	0	0
2	0	1	1	0
3	1	0	1	0
4	1	1	0	1

arithmetic operation might require the adders to complete several dozen additions and shifts before an answer is displayed. Obviously, if the same adders are used for all of these computations they can not be done at the same time. They have to be done serially, and each computation must be "clocked" in and out of the adders at a precise time. There cannot be any overlap, because different sets of data arriving at the inputs of the adders at the same time will result in errors.

Clocks

Special timing pulses called clocks are used to establish the proper timing in digital circuits. A single clock pulse is often used to time hundreds of different operations in a digital computer. There are many ways to generate a clock pulse. Some circuits generate their own, and do not need an external clock. These circuits may have special clock-generation sections called ring counters or up/down counters. Circuits that do not generate their own clocks have to depend on an external multivibrator circuit for their timing.

Multivibrators

A multivibrator is a digital oscillator which produces a serial stream of pulses, or waves. In radio, and other non-digital oscillators, the output is usually a sine wave that varies in both frequency and amplitude. In digital circuits, the output is a square wave. The frequency is predetermined and constant, and each square wave pulse is the same amplitude (size) as its neighbors. Figure 2-39 illustrates both types of oscillator outputs. The square wave is used in digital electronics because the transition from "low" to "high" or "high" to "low" must occur quickly, and the "more square" the clock pulse is, the faster the digital gates and other logic circuits are able to change states. The two most common types of multivibrators in digital electronics are the *astable* and *monostable*.

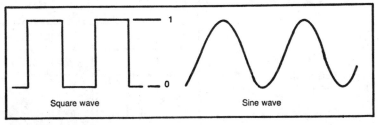

Fig. 2-39. Oscillator waveforms.

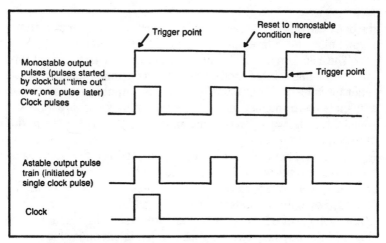

Fig. 2-40. Astable and monostable multivibrator outputs.

Astable Multivibrator

An astable multivibrator is one that, when triggered, supplies a continuous stream of square waves. The name "astable" comes from the fact that once this type of multivibrator starts running, it cannot stay in one state. When the output goes high, it immediately wants to go low, and when it goes low, it wants to go high. The output is not stable in either the high or low state. The output frequency and pulse width is usually adjustable, making astable multivibrators extremely accurate. Digital clocks often use a type of astable multivibrator, and they are often accurate to within a second per month. The output waveform of an astable multivibrator is shown in Fig. 2-40.

Monostable Multivibrator

Monostable multivibrators are sometimes called *one-shot*, or *single-shot*. They are stable in only one state, usually low. When triggered, a monostable multivibrator will supply a single high square-wave pulse. Between triggers, the output remains low. The pulse width of the output square wave is usually adjustable. The typical monostable output waveform is shown in Fig. 2-40.

555-Type Industrial Timer

The 555-type industrial timer, in spite of its name, is actually a monostable/astable multivibrator. It is simple-to-use, versatile, inexpensive, and readily available. Some of the exercises later in

39

the book use this circuit, and since the 555 is a multivibrator, this is a good place for a brief description of its operation.

The 555 is not a TTL integrated circuit. It is a related type known as a linear integrated circuit, and its primary function is as a timer for industrial controls. However, its output is a square wave, and since the frequency and pulse width of these square waves is adjustable, the 555 makes a fine multivibrator. The time period between output pulses can be as short as a few microseconds or as long as several hours. The 555 can be operated as either a monostable or astable multivibrator. In the astable mode, it will maintain an accurately controlled, free-running frequency and pulse width determined by only two external resistors and one capacitor. With a supply voltage of +5 volts, as is most often used, the output voltage (high) of the 555 is about 3.5 volts, which is excellent for driving TTL circuits.

In the exercises later in this book, the 555 will be used to supply square waves at various frequencies and with various pulse widths. These variables are controlled by changing the values of the external resistors and capacitors, and this process is clearly explained and illustrated in each exercise where the changes are necessary.

FLIP-FLOPS AND SHIFT REGISTERS

A flip-flop is a data-storage device, and as such it is used quite extensively in counters, shift registers, and simple memories. The simplest flip-flops are constructed using only two NAND gates. Shift registers consist of a number of flip-flops connected serially, and data are "shifted" in one end of the register and out the other. There are different types of shift registers, but in the basic type the first data in is the first data out. The period of time spent by the data in the register is the storage time. This unit will explain how a flip-flop works and how it is used in a shift register.

There are several types of flip-flops, and they are usually labeled according to the way in which they operate. The most common types of flip-flops are the RS (Reset-Set), D, and J-K. The most basic of these is the RS type.

RS Flip-Flop

Figure 2-41 shows the logic diagram of an RS flip-flop. In this circuit (as in all flip-flops) there are two outputs, Q and $\overline{Q}$. They are complements of each other. This means that if Q = 1, then $\overline{Q}$ = 0, and if Q = 0, then $\overline{Q}$ = 1.

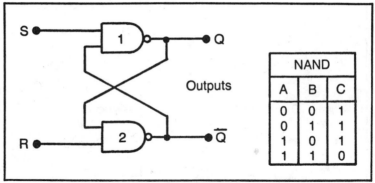

Fig. 2-41. Basic RS flip-flop block diagram.

Suppose the S (set) terminal of the flip-flop in Fig. 2-41 were to be made low (0), and R (reset) made high (1). Suppose Q (output) is also high (1). NAND-gate 1 then would be "on" since both inputs to it would be low. See the truth table included in Fig. 2-26 to verify this. The Q output is tied to one input of NAND-gate 2 so this input is high (1). The other input to NAND-gate 2 is also high (1). This makes the output of NAND-gate 2 low (0). This is a stable condition with $Q = 1$ and $\overline{Q} = 0$. If the R input were changed to a low (0) and the S input to a high, gate 2 would cause $\overline{Q}$ to change to a high (1) condition. This would be communicated to the input of gate 1 and it would change its output to a low (0) condition. The output from each gate is run back to the input of the other gate. This prevents the two gates from having the same logic-state output at the same time.

To summarize, if S (set) is low, the RS flip-flop will have an output of $Q = 1$, $\overline{Q} = 0$. If R (reset) is low, the output will be $Q = 0$, $\overline{Q} = 1$. It is important that *both* R and S not be low at the same time. If this is done, the output logic states cannot be controlled.

D Type

The D type flip-flop has the same characteristics as the RS type, but it also has a capability that the RS type does not have. Figure 2-42 shows the block diagram of the D flip-flop. The D flip-flop is a "clocked'" unit. This means that the outputs (Q and $\overline{Q}$) are controlled by an input clock (square wave) of some frequency. The logic placed on the "D" input is transferred to Q on the next clock pulse. This transfer occurs when the clock signal goes from 0 to 1 (low to high). This is known as the "leading edge" of the clock signal. In some cases, the "trailing edge" of a clock signal (when the logic goes from 1 to 0) will activate a unit. Both types are available.

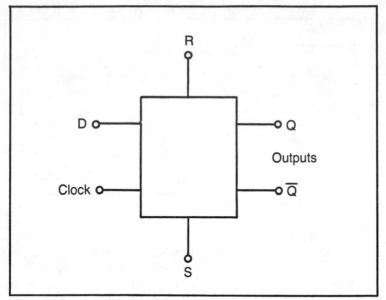

Fig. 2-42. D type flip-flop using 2-input NAND gates.

The D type flip-flop is useful in applications where a bit of data is transferred from D to Q, or held in memory.

A latch is another type of circuit that is sometimes made from D flip-flops. In a latch, data is sometimes shifted through the flip-flop at high speed. In these cases the data is applied to D and exits at Q. If either the D connection or the clock signal is interrupted, the output at Q will remain in the logic state of the last bit that arrived before the interruption. An example of this process is a digital stopwatch. Usually, a number of flip-flops are used to handle the flow of data. For example, if the data is grouped into four-bit words, four flip-flops can be used to "latch" the four bits of the binary data words. In a stopwatch, the clock keeps on counting time and only the display is "latched" in order to read a lap time.

J-K Types

The J-K flip-flop has capabilities that make it even more useful than the RS or D types. Figure 2-43 shows the block diagram of the J-K flip-flop, sometimes called a master-slave flip-flop. The J-K has the set and reset controls found on the other types, and it is also a "clocked" unit since a clock signal controls the flip-flop of the output logic. The J and K terminals are used to "steer" the input logic. The following rules apply to a typical J-K flip-flop:

42

1. If either R or S is low, the unit will "clear" to Q = 0, $\overline{Q}$ = 1 (reset), or Q = 1, $\overline{Q}$ = 0 (set). R and S must not be low at the same time or output control will be lost.

2. If J and K are both made high (1), Q and $\overline{Q}$ will divide the clock frequency by 2.

3. If J and K are both made low (0), the Q and $\overline{Q}$ outputs will do nothing. This is an easy way to disable the unit.

4. If J = 1 and K = 0, the next clock pulse will make the outputs go to Q = 1, $\overline{Q}$ = 0.

5. If J = 0 and K = 1, the next clock pulse will cause the outputs to go to Q = 0 and $\overline{Q}$ = 1.

Figure 2-44 shows the diagram for the J-K flip flop along with an illustration of the logic operation of the output when J and K are placed at various logic levels.

The major use of the J-K flip-flop is in complicated applications such as shift registers. A wide variety of logic "steering" situations can be handled with J-K flip-flops.

Shift Registers

Shift registers are made of a number of flip-flops. These flip-flops pass digital data from one to another, and this "shift" of data is where the name comes from. This type of circuit can store and read out data in both *serial* and *parallel* forms. The shift register can do

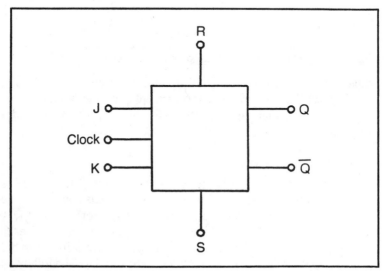

Fig. 2-43. J-K type flip-flop block diagram.

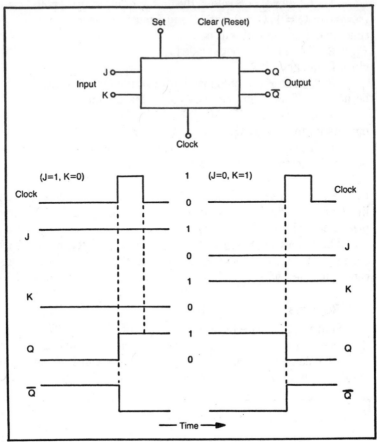

Fig. 2-44. Logic timing showing leading-edge operation of J-K flip-flop.

other operations, such as digital up and down counting. The J-K flip-flop will be used to show the principles of operation of shift-registers. A special register, the ring counter, will also be described.

Four-Stage Shift Register

In Fig. 2-45, a shift register consisting of four J-K flip-flops is shown. At the "A" flip-flop, for purposes of discussion, J is made high (1) and K low (0). All flip-flops are cleared (reset) to $Q = 0$, $\overline{Q} = 1$. The output will be monitored only at the Q output of each flip-flop. The Q and $\overline{Q}$ outputs of each flip-flop are tied directly to the J and K inputs of the next flip-flop (A to B to C to D). The following steps will trace the progress of data through this circuit:

1. Enter a bit of data (1) into flip-flop A by using the "set" button switch. The register now should read the binary word

$$A\ B\ C\ D$$
$$1\ 0\ 0\ 0$$

2. The Q and $\overline{Q}$ of flip-flop A now make J = 1 and K = 0 of flip-flop B.

3. The next pulse from the clock causes flip-flops A and B to flip-flop. The others have J = 0, K = 1 and so remain in the same state, Q = 0, $\overline{Q}$ = 1. This means that the digital word has now changed to

$$A\ B\ C\ D$$
$$0\ 1\ 0\ 0$$

4. The next pulse then will cause the 1 in flip-flop B to go to 0 and the 0 in C to go to a 1. The word is now

$$A\ B\ C\ D$$
$$0\ 0\ 1\ 0$$

5. The next pulse will cause the word to change again to

$$A\ B\ C\ D$$
$$0\ 0\ 0\ 1$$

6. And the last pulse will shift the 1 in D out so that the word now is

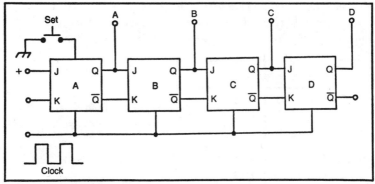

Fig. 2-45. The 4-stage shift register diagram.

Figure 2-46 shows the progress of the bit through the register as it is "shifted" from left to right. This left-to-right shift also effectively performs the function of division. Figure 2-47 shows this dividing effect. Notice that each time a bit is shifted one place, the decimal equivalent is divided by 2.

If the register is set up so that the shift is from right-to-left, the result will be a multiplication by 2.

In the example shown in Fig. 2-46, the data is entered in a serial form (one bit after another) and shifted out in parallel form (all bits can be read at the same time). Both forms of data entry and readout are used with shift registers. Data can be entered either in serial or parallel form and read out in either form. It all depends on how the register is designed.

Ring Counter

In the shift register shown in Fig. 2-45, the data is shifted out of the register after four pulses and is lost. The only way data can be entered is through the "set" control on the flip-flops. This register can only "count" or show 5 different states- and only once at that. The *ring counter* is a shift register that re-circulates the bits of information, thereby repeating the counting process over and over again.

Figure 2-48 shows the block diagram of a four-stage ring-counter shift register. In this case, the data bit is entered into flip-flop A and is passed on, left-to-right, with each succeeding pulse of the clock. However, notice that the Q and $\overline{Q}$ outputs of the D (last) flip-flop are hooked to the J-K inputs of flip-flop A. This puts the bit back into A and keeps it going through the register over and over again in a "ring." This counter makes a complete cycle every 5 pulses of the clock. This type of register can be used to count

	A	B	C	D
Beginning	0	0	0	0
Set A	1	0	0	0
Pulse 1	0	1	0	0
Pulse 2	0	0	1	0
Pulse 3	0	0	0	1
Pulse 4	0	0	0	0

Fig. 2-46. Truth table showing shift of a bit through the register.

Step	Digital word				Decimal Equivalent
	A	B	C	D	
1	1	0	0	0	8
2	0	1	0	0	4
3	0	0	1	0	2
4	0	0	0	1	1

Fig. 2-47. Truth table showing left-to-right shift that produces division.

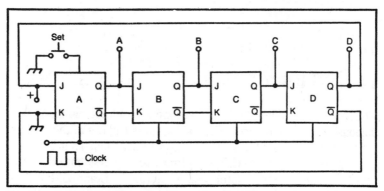

Fig. 2-48. 4-stage ring counter shift register diagram.

pulses, numbers, or any other digital event. Any one of the outputs, say D, will represent a count of 5 each time it goes high (1) making this a divide-by-five counter.

The process of counting is very important in digital circuits. The basis of counters is usually a flip-flop shift register such as the ring counter. By adding other gates, these counters can be modified to count in terms of nearly any number. Some commonly used counters, such as divide-by-10 and divide-by-2 (a simple flip-flop) are used so often that special integrated circuits have been developed specifically for these jobs.

Chapter 3

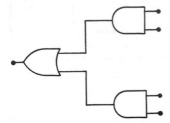

Logic Functions

This section is devoted to the development of logic functions with various solid state devices. First it explores the use of the transistor for switching and for driving high-current loads. Then I explain the light-emitting diode (LED) and its use as a readout device for digital circuits. This chapter also explores TTL and CMOS, the most common logic families, and provides a practical explanation of integrated circuits and how they are used.

BIPOLAR SWITCHING TRANSISTORS

The transistor, when used in digital applications, usually functions as a switch. It is more commonly known as an amplifier in such applications as radios and hi-fi amplifiers. The type of transistor used for switching must be able to switch states (on-to-off or off-to-on) very rapidly in order to handle the digital data that is applied to it. This unit will show how typical PNP and NPN transistors are used to drive digital loads and switch digital pulses.

The Bipolar Transistor

There are a number of types of transistors. The one most widely used in digital applications is the bipolar type. This transistor is known for its fast switching speed. Transistors of this sort come in a variety of case styles. Figure 3-1 shows a number of package types that are used for different kinds of transistors. Figure 3-2 shows the schematic symbols used for the two basic types of

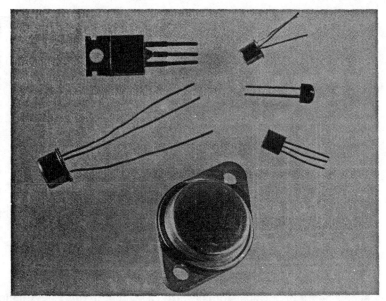

Fig. 3-1. Variety of transistor case styles. Large unit at bottom is a power transistor (photograph by Mary Duncan, Oswego Learning Resources Center.)

transistors, the PNP and NPN. In both cases, the *BASE* is used as the input, or control, elements. The signal on the base determines the state of the current (on or off) through the emitter - collector connections.

Switching

In many cases, switching can be done with gates such as NAND, NOR, or by inverters or buffers. If a gate is not available, a transistor can be used. Usually, a gate is preferred because it does not use as much current as most bipolar transistors. Figure 3-3

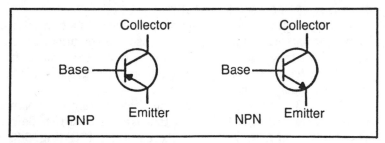

Fig. 3-2. Schematic symbols of PNP and NPN transistors. The arrow is always attached to the emitter connection.

49

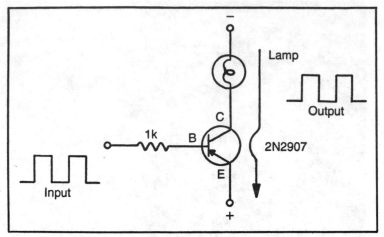

Fig. 3-3. PNP switching transistor. Curved arrow indicates current flow when the transistor is turned "on."

shows how a PNP transistor is used as a switch. In this case a lamp is used to indicate the output, so the transistor must supply a relatively high current (40 milliamps or so). This amount of current is higher than can normally be supplied by a gate.

In Fig. 3-3, if no signal (0 logic) is present at the input, the transistor is off and no current flows from the negative (−) side of the power supply to the lamp through the E (emitter) and C (collector) and on to the positive (+) side of the supply. When a positive (+) (1 logic level) is applied to the input, the transistor is "turned on," allowing current to flow through the E-C path to the lamp. The lamp, in this case, is the load and limits the amount of current flow. In some cases, a resistor may be placed in series with the lamp in order to further limit the current. This is often done if the supply voltage is much higher than the lamp rating. It is a must when the lamp is a light-emitting diode (LED).

In Fig. 3-4, another example of switching with a transistor is shown. In this case, the transistor is used to condition, or "debounce," the toggle switch. In practice the transistor is most often used to switch an output and to drive high-current loads. It is not used very often to debounce a switch.

Figure 3-5 shows the switching application using an NPN type transistor. Notice that in the NPN transistor, the current flow is from the emitter *to* the collector, whereas in the PNP (Fig. 3-3) it was from the collector *to* the emitter. The major difference between the two is that the *resting logic* level is opposite in the two cases. In

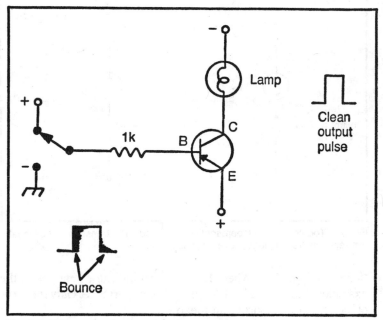

Fig. 3-4. Using a transistor to "de-bounce" a toggle switch.

the PNP case, the resting logic is low (logic 0), and in the NPN the resting logic is high (logic 1). Figure 3-6 illustrates these two conditions. The transistors have been replaced with simple switches because that is their function. The "resting" logic level becomes important if the output signal is to be passed on to addi-

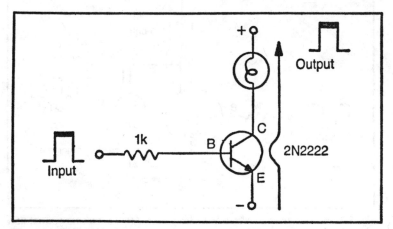

Fig. 3-5. NPN transistor used for switching. The arrow indicates current flow through the transistor when it is "on."

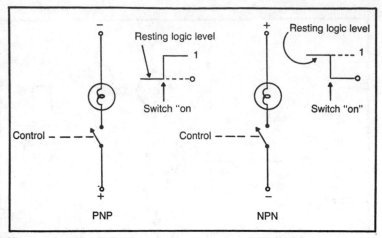

Fig. 3-6. Toggle switches representing PNP and NPN transistors showing the difference in "resting logic" level for each.

tional logic devices. When they drive an indicator lamp, as in the examples, the resting logic level is of no importance. Only the on or off condition is important in these cases.

Inverters

A transistor can be used to invert a logic signal or state. Figure 3-7 shows the wiring for a PNP transistor inverter. When the logic at the input is low (0), the transistor is "off" and no current flows

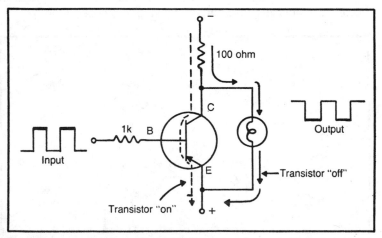

Fig. 3-7. Inverter function with a PNP transistor. Arrows indicate current flow when the transistor is "on" and "off."

through the transistor. The current can, however, flow through the 100-ohm resistor and the lamp. The lamp is then "on" or in a high (1) logic state. This is the inverse of the input. When a high state (1) is applied to the input, the transistor turns "on." Current will now pass through the 100-ohm resistor and the transistor since this path has less resistance than the resistor-lamp path. Remember, the transistor, when it is on, is like a closed switch, and it has little or no resistance. When the transistor turns on, the lamp will turn off. Again, the output is an inversion of the input. Figure 3-8 shows the same inverter function using an NPN transistor.

In most cases where a transistor is used to switch or invert, the common 2N2222 (NPN) or 2N2907 (PNP) will do well. Each of these transistors will deliver about 800 milliamps (0.8 amps), which is sufficient to light most display lamps. Each will handle voltages (collector-to-emitter) of up to about 30 volts and frequencies of up to 350 megahertz.

There are some differences between the two in limitations, so consult a transistor substitution guide for the exact rating of any transistor used.

Although the outputs of the examples shown in this unit are lamps, other devices can also be powered by the transistor. Relay coils, reed relays, LED indicators, SCRs and TRIACs are a few of the devices that are commonly driven by transistors.

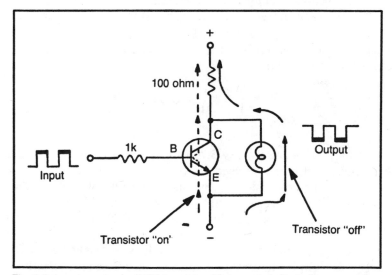

Fig. 3-8. Inverter using an NPN transistor. The arrows indicate the current flow when the transistor is "on" and "off."

READOUT DEVICES

Since most digital circuits use the binary number system for internal operation, an easy means of entering and reading-out this "language" is necessary. Entering information is usually done with keyboards of various kinds. The keys automatically produce a series of digital pulses that represent the binary word. The readout is somewhat different. Readouts can be simple light-emitting diode (LED) indicator lamps, elaborate printers, or television-screen displays. This unit will explore the use of the LED (light-emitting diode) and the seven-segment display.

LED

The LED (light-emitting diode) is a diode that emits light when current passes through it in the forward (cathode to anode) direction. The symbol for the LED is shown in Fig. 3-9. This illustration also shows two types of LEDs and how to identify the cathode lead.

As the current through the LED increases in the forward direction, the light intensity also increases. A LED, however, can be easily destroyed unless a limiting resistor is placed in series with it. The LED must be placed in a circuit so that current flows in the forward direction (cathode to anode), If it is placed in the circuit in the reverse direction, no light will be emitted. Figure 3-10 shows the correct hook-up for proper operation.

The value of the series limiting resistor used with most LEDs depends on the applied voltage. Most LEDs operate at a current of approximately 10 milliamps at 5 volts dc. Ohm's law indicates a resistance value of:

$$R = \frac{E}{I} = \frac{5}{.01} = 500 \text{ ohms } (\Omega)$$

In most applications where a 5-volt source is used to light an LED, a 330-ohm resistor can be used. If the voltage is higher, say 12 volts, then the limiting resistor value should be raised to about 1000 ohms. *Never* try to operate an LED without a limiting series resistor. The LED can be ruined instantly if the limiting resistor is not used.

LEDs are available in red, green, yellow, and clear colors. The red color is by far the most popular. The other colors are sometimes hard to see, especially in bright daylight. LEDs can be purchased in a variety of sizes from a pin-point up to large devices ½ inch in

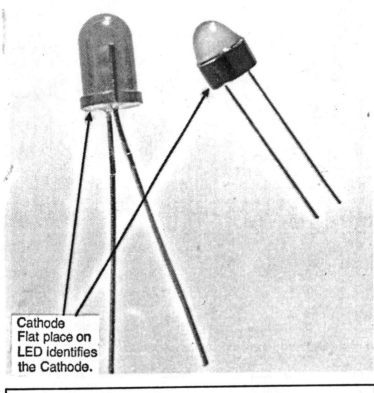

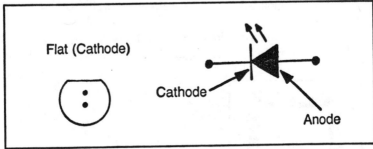

Cathode
Flat place on
LED identifies
the Cathode.

Flat (Cathode)

Cathode

Anode

Fig. 3-9. Symbol (left) of LED and illustration of how the cathode is identified. (photograph by Mary Duncan, Oswego Learning Resources Center.)

diameter. The larger units use a plastic magnifying lens to diffuse and distribute the light over a larger area. Most LEDs are enclosed in a plastic case that is the color of the emitted light.

Seven-Segment Readout

Another very useful readout is the seven-segment LED read-

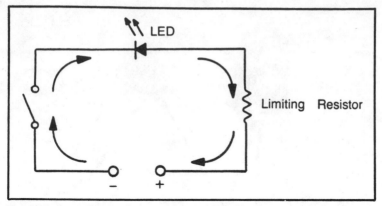

Fig. 3-10. Correct wiring for operating an LED. The limiting resistor value depends on the applied voltage. Arrows indicate direction of forward electron flow.

out. Figure 3-11 shows the layout of this type of device. The seven segments can be activated separately or together in any desired combination. Each segment consists of one or more LEDs. Frequently the LEDs are covered by a "bar" of colored plastic that "glows" when the LEDs underneath are "on." By selecting the correct segments, all the numbers 0 to 9 can be represented, and a number of letters of the alphabet can also be displayed. This display can provide a direct readout in numbers (such as a calculator display) that is far superior to a readout in binary code. Figure 3-12 provides a comparison of the seven-segment, BCD and decimal numbers 0 to 9.

Seven-segment readouts come in two major types; common anode and common cathode. Figure 3-13 shows the common-anode

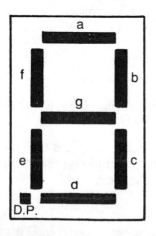

Fig. 3-11. Seven-segment LED readout.

Decimal	BCD				Seven-Segment						
	D	C	B	A	a	b	c	d	e	f	g
0	0	0	0	0	1	1	1	1	1	1	0
1	0	0	0	1	0	1	1	0	0	0	0
2	0	0	1	0	1	1	0	1	1	0	1
3	0	0	1	1	1	1	1	1	0	0	1
4	0	1	0	0	0	1	1	0	0	1	1
5	0	1	0	1	1	0	1	1	0	1	1
6	0	1	1	0	0	0	1	1	1	1	1
7	0	1	1	1	1	1	1	0	0	0	0
8	1	0	0	0	1	1	1	1	1	1	1
9	1	0	0	1	1	1	1	0	0	1	1

0 = Low (off)
1 = High (on)

```
 _a_
f|_g_|b
e|   |c
  d
```

Fig. 3-12. Truth table showing equivalents in decimal, BCD, and seven-segment forms.

arrangement and Fig. 3-14 shows the common-cathode type. In one case, the anodes are all tied together (common anode) to the positive (+) side of the power supply. A limiting resistor is placed in series with *each* LED segment. When a logic low (−) is applied to the input or resistor, that LED will light. In the common-cathode type, the cathodes are all tied together to the negative (−) side of

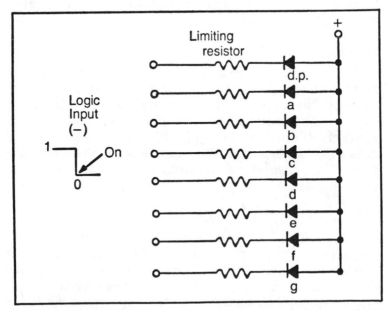

Fig. 3-13. Common-anode seven-segment display schematic. Resistors shown are usually external to the seven-segment display unit.

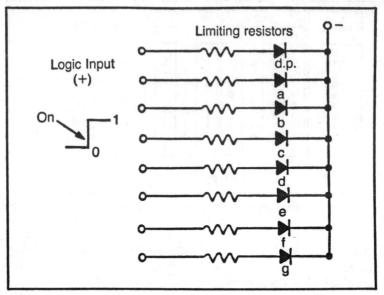

Fig. 3-14. Common-cathode seven-segment display diagram. Limiting resistors are usually external to the LED display package.

the power supply and each segment is activated with a logic high (+) through a limiting resistor.

Other Devices

There are other devices used to display information: nixie tubes, neon-type tubes, and fluorescent readouts are also available. The LCD (liquid crystal display) is becoming popular because it uses very little current. There are also devices that consist of arrays of LEDs designed to display alphanumeric symbols. These units are usually expensive. When there is a need to display many letters of the alphabet, the usual solution is a video screen or a teleprinter device.

INTEGRATED-CIRCUIT LOGIC FAMILIES

Digital-logic operations are usually done electronically with *integrated circuits*. The various families of integrated circuits have their own characteristics, advantages, and disadvantages. At the present state of the art, the two most prominent integrated circuit families are the TTL and CMOS. These two comprise the bulk of the ICs now in use. The TTL family is the more popular, but CMOS is quickly gaining ground. This section will explain the characteristics of both TTL and CMOS as well as the size and character of the various scales of integration.

Early Types

The first integrated circuits (ICs) were developed in order to reduce the size of the finished circuit. The number of vacuum tubes was reduced by placing several tubes in the same glass envelope. Only one socket was required, substantially reducing the amount of wiring in a circuit.

In integrated circuits, the same principle was used to put many transistors on the same "chip" of solid-state material. Silicon was usually the semiconductor material used. Other components such as diodes, resistors (adding resistive material paths), and capacitors (conductive layers separated by insulative layers) were soon added to the single chip. This process allows a complete circuit, such as a radio receiver, to be manufactured on a single tiny chip that is encased in a single package.

The type of logic was named after components that were used to do the switching. The early logic type was the DTL (diode-transistor logic). In this type of IC, diodes and transistors provided gates and other functions. Another early logic type was the RTL (resistor-transistor logic) family.

TTL

At present, the most widely used integrated circuit logic family is the TTL (transistor-transistor logic) family. The 5400/7400 series of ICs that are used in all the examples and exercises in this book are TTL types. A wide variety of TTL ICs for nearly all logic applications is available at low cost. These devices are available in high and low power as well as high-speed types.

CMOS

The CMOS (complementary metal-oxide semiconductor) family is growing very rapidly and will probably become the dominant digital IC family in the near future. CMOS offers many advantages over the TTL types.

The greatest advantage of the CMOS IC is that it uses very little power in comparison to TTL devices. The CMOS IC can be constructed on smaller chips because the density (number of components per unit of area) can be much higher than with TTL. The one disadvantage of CMOS is that it is slower, generally, than TTL.

This may not be important if the applications do not require very high switching speeds.

Early CMOS units were sensitive to static discharge, and they had to be handled with extreme care. Newer units have added internal input-protection circuitry to overcome this disadvantage, and they can be handled without any special precautions.

There are CMOS ICs available that perform nearly all logic functions. They are usually encased in the DIP (dual-in-line package). The cost of CMOS ICs is a little more than TTL, but this is offset by the fact that fewer CMOS units are needed to perform many functions. For example, a decade counter made with TTL devices frequently requires a counter (7490), a driver (7447), and possibly a latch (7475). A single CMOS unit (4026) will replace both the counter and driver, and it uses much less power than the equivalent TTL circuitry.

Integration Scales

As more and more integrated circuits were designed and manufactured, the trend was to reduce the size of the "chip" by packing more components into a smaller area (increasing density). There are a number of scales used in the industry to place chips in a general category according to the number of components per unit of area. The basic unit of measurement is usually the logic gate.

SSI (small-scale integration) is the term applied to the smallest of the integrated circuits. They generally perform a single function such as a gate or a flip-flop. The SSI IC usually contains less than 12 gates or their equivalent.

MSI (medium-scale integration) devices usually contain a complete circuit. An MSI unit will contain between 12 and 100 gates or their equivalent. A digital counter or driver is usually an MSI unit.

LSI (large-scale integration) devices contain at least 100 gates or their equivalent circuitry. Complete functional devices such as calculators and digital clocks are usually contained on LSI chips. Figure 3-15 shows a complete calculator on one integrated circuit. LSI devices with as many as 100,000 transistors on a single chip are commercially available.

VLSI (very-large-scale integration) devices are now being built, and they typically contain entire circuits such as computers. They are frequently used to provide low-space-consuming memories for mini-and micro-computers. The VLSI chip contains at least 1000 gates or their equivalent circuitry.

60

Fig. 3-15. A complete calculator contained on one "chip" or LSI device. A cover normally protects the insides. Notice that small wires connect the chip to the pins of the holder. The actual chip is approximately ¼" square (photograph by Mary Duncan, Oswego Learning Resources Center.)

The present technology uses photographic processes to produce the very fine detail necessary in integrated circuits. The VLSI chip is so high in density that normal light processes do not yield good results. Present experimentation is moving beyond the use of visible light to the use of x-rays and lasers. New processes are constantly being found and developed, and they hold the promise of even greater degrees of miniaturization.

TTL INTEGRATED CIRCUITS

In this section, the characteristics of the TTL integrated circuit will be discussed, and the practical side of using TTL ICs will be explored with information about the speed, electrical characteristics, and external connections.

Dip Package

Nearly all TTL integrated circuits are contained in the DIP (dual-in-line package). These packages have 14, 16, or more pins or contacts in two rows. Figure 3-16 shows the typical outline and numbering sequence of the DIP package. These packages are also used for other logic families such as CMOS.

In order to locate pin 1, an identification notch or dot (or both) is located at one end of the package. Pin 1 is to the left (from the top view) of the notch or near the dot. Pins are counted (or numbered) starting from pin 1 and proceeding in a counter-clockwise direction. A number of books and substitution guides are available which show the pin-out diagrams of common ICs. Manufacturers usually provide

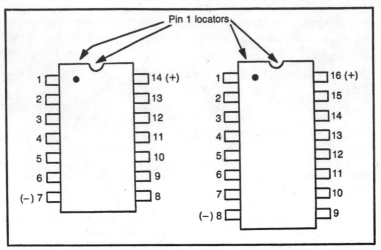

Fig. 3-16. DIP packages for 14 and 16 pin types. View is from the top of the unit. The bottom pin on the left (7 or 8) is usually the negative supply input, and the top pin on the right (14 or 16) is usually the positive supply input.

this information when units are packaged individually, such as in a blister pack.

Power

TTL integrated circuits usually require a 5-volt supply. This voltage must be regulated so that little variation in voltage occurs when changes in current are demanded. In an unregulated power supply, each time the current demand changes, the voltage also changes, and sometimes this voltage change is great. With digital ICs, a voltage change can cause the IC to function improperly. The result is often an error in the logic or in the resulting output. Solid-state regulators are available that are easy to use and inexpensive. Figure 3-17 shows the schematic of a typical power supply used for TTL IC work. A 5-volt regulator is used, and the transformer must be able to deliver enough current for the desired output. The capacitors and rectifier must also be selected for the appropriate current and voltage. A 7805 regulator IC will provide 5 volts dc at 750 milliamps. This is sufficient for most TTL applications. If an LM309K regulator is used, a maximum of 1 ampere can be obtained.

When designing a power supply for TTL, always determine the current requirements of each IC and add them all together (the total number of ICs) to determine the total current needed. Always design the power supply so that *more* than enough current is available. Information about the current requirements of particular ICs can be obtained from manuals such as "The TTL Data Book" published by Texas Instruments.

Fan-In and Fan-Out

Fan-in refers to the number of inputs that can be delivered to the IC in use. Fan-out refers to the number of units that can be *driven* by the output of an IC. The typical fan-out for a TTL unit is 10. This means that 10 TTL ICs can be driven by the single output of another TTL IC. The fan-out of CMOS is much greater than TTL—as many as 50 units. This is due to the very low input drive required for CMOS ICs.

Fan-in is usually a single signal, since each input usually has its own source of drive. If many inputs are to be delivered to one IC, arrangements must be made to prevent these inputs from being mixed and lost.

Switch Conditioning

In most TTL ICs, the input signal must be a square wave. Flip-flops, counters, and shift registers will frequently "count" any pulse that is presented to them, even false pulses. Very short pulses will be counted because these units react very quickly. Mechanical switches are one source of false pulses. Figure 3-18 shows the "bouncing" that occurs when a mechanical switch closes or opens. Also shown is the type of circuit used to eliminate this bounce. De-bouncing, or conditioning, of the signal is necessary in certain TTL circuits when mechanical switches are used.

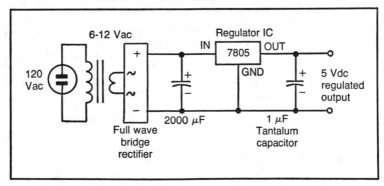

Fig. 3-17. Schematic of a 5-volt regulated dc power supply. Either a 7805 (750mA) or 309K (1A) regulator IC can be used.

False pulses may also be caused by a heavy switching demand by a number of ICs. These pulses are usually called "glitches." A glitch is a pulse of unknown origin—usually unwanted. The use of bypass capacitors at the positive power connections of every fifth IC or so usually eliminates this sort of problem. Bypass capacitors should be of low values, say .01 μf.

Readouts

The logic-state readout of a TTL IC is easily done with a LED. In cases where switching is occuring slowly—say less than 1 hertz—the LED (with a series-limiting resistor of 330 ohms) can be

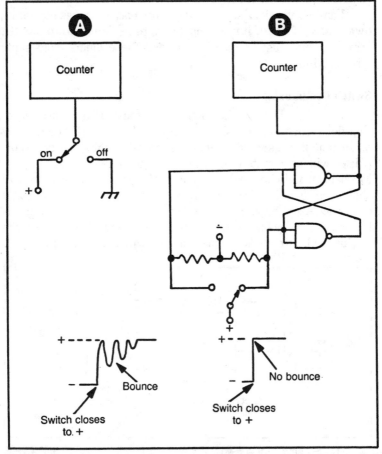

Fig. 3-18. Circuit (a) shows the bouncing of the mechanical switch. The circuit at (b) eliminates the bounce.

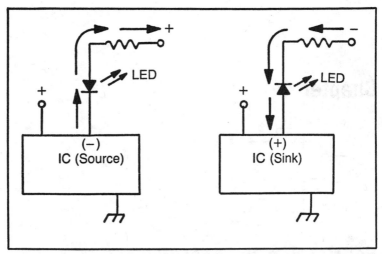

Fig. 3-19. Integrated circuit shown as both a source (supply) of current and a sink (user) of current. Arrows indicate current flow in each case.

permanently mounted. In this way a continual visual check of the circuit operation is possible.

If numerical outputs are desired, then a driver IC and a seven-segment readout is probably the best solution. Most LED readouts will work well on the 5-volt output of TTL ICs. In some cases, the current comes *from* the IC (this is known as a *sourcing* IC) and in others the current must go *into* the IC. This latter system is called *sinking* because the current must "sink" into the IC. Figure 3-19 shows a typical example of both *source* and *sink* applications. In each case, the IC provides the switching. The LED must be inserted in the correct polarity so that the circuit will function properly. Notice that the LEDs are reversed in the illustrations in Fig. 3-19.

If a large amount of current or voltage is needed to operate a circuit, it may be necessary to drive a transistor which, in turn, will drive the circuit. In this application, the action of the transistor is similar to that of a relay. Always check the output current limitations of the IC in use. If the unit cannot source (or sink) the required current, then a transistor may be the best solution.

Chapter 4

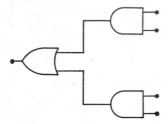

Fabrication Process

This section covers working with integrated circuits and printed circuit (PC) boards. A brief treatment of PC board design and fabrication is given followed by practical information about working with breadboards, wirewrapping, power supplies, and other details of IC operation.

Complete plans for an experiment board are provided. This experiment or exercise breadboard can be used to wire all of the exercises in Chapters 5 through 10. For more detailed coverage of the fabrication of PC boards, see *Digital Electronics Projects* by Harry Hawkins (TAB Book No. 1431).

PRINTED CIRCUIT BOARDS

Printed circuit (PC) boards are widely used in present day electronics. This chapter provides a brief explanation of how these boards are designed and fabricated. Soldering, etching, resist application and drilling are covered.

PC boards are usually designed on paper before being fabricated on a copper-clad laminated-plastic sheet. The circuit is first breadboarded or wired by hand to verify that it indeed operates as desired. The completed design is then transferred to the plastic board, and copper paths replace the wires as conductors for the circuit. The component parts are mounted to the board through drilled holes, and the wires from each component are then soldered to the copper. The finished board serves the dual purpose of holding the components and forming part of the circuit.

Design

PC boards are designed on paper in order to develop the smallest area necessary for the components. Figure 4-1 shows the schematic diagram of a circuit that is to be fabricated on a PC board. The electrical values of the components (resistors and capacitors) are not really important for this example, but their physical sizes must be known in order to determine how they will fit in with the other parts on the board. It is good practice to use the *actual* components so that exact clearances will be known.

The next step in the design of a PC board is the first layout. This often duplicates, to some extent, the schematic diagram, at least in general form. Figure 4-2 shows one "first try," or design. This first design can be improved some by moving components around in such a way that maximum use is made of the smallest area. It is a good idea to avoid "jumpers" or wires used to bridge over a conductor path. Jumpers require time to install and so should be avoided unless necessary, but there are times when a jumper is the only practical solution. Try to arrange all incoming and outgoing connections to the PC board around the edges of the board. Avoid running wires into the center of the board. This simplifies troubleshooting later on.

Figure 4-3 is an improved version of the design in Fig. 4-2. Notice that *both* sides of the PC board are shown. It is important to

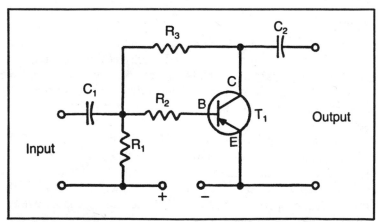

Fig. 4-1. Schematic diagram of circuit to be placed on a PC board.

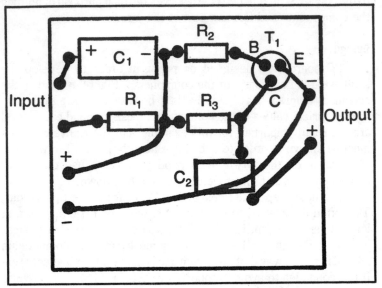

Fig. 4-2. "First try" design in Fig. 4-1. Lines indicate the foil side where the copper paths will be needed. View is from the component side. Components are approximately full scale.

make sure the *foil side* (copper-clad side) is correct since it is the one that will be used to make the final PC board layout. The beginner will usually make about three tries before a good design is obtained. An experienced designer can make a good board on the first try.

Resist

Once the correct *foil-side* layout has been designed, it should be rechecked using a red pencil and a copy of the schematic diagram. It is important to complete this check. If any mistakes have been made it is much better to find out at this point rather than after a board is finished. If no errors are found, then the resist may be applied.

A *resist* is an acid-resistant material that is applied to the copper according to the foil pattern layout. The resist material will prevent the etching solution from coming into contact with the copper it covers. This "protected" copper will remain after the unprotected copper is etched away. This remaining copper eventually becomes the conductive paths for the circuit.

Many materials can be used for resist purposes. Most laquers and enamels, finger-nail polish, and some tapes, will do an acceptable job as a resist. The resist pen, available at most local electronics stores, is an excellent device for applying resist.

First, the copper must be cleaned with a fine steel wool. Then the full-scale pattern of the foil side is laid over the surface of the copper. An awl is used to press small marks through the paper layout onto the copper at each place (pad) where a hole is to be drilled. The layout paper is then removed and used as a guide to connect the "pads" with the resist material. The result is a duplication of the foil layout drawing on the copper with a resist material. The resist must be dry before the etching process is started. In some cases, the resist dries almost immediately, and in other cases it may have to be baked or heated.

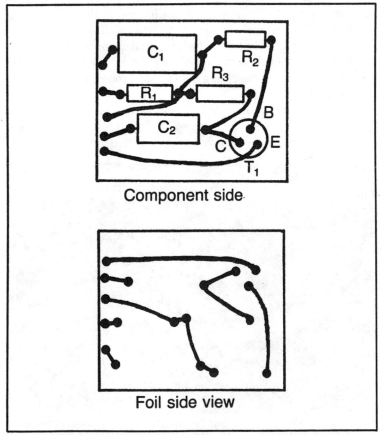

Fig. 4-3. Final design of PC board showing both component and foil side layouts. The actual PC board will be made from the foil side view layout. These layouts are approximately full scale.

Etching

Etching is the removal of unwanted copper foil from the PC board. Usually, a mild acid is applied to the unprotected copper foil. This may be done by immersing the board in a bath of *etchant*. The etchant (or the work) is kept in motion so that fresh etchant is always in contact with the copper. The etchant can also be sprayed on the copper. In both cases, the temperature of the etchant should be raised to about 100° F to speed up the chemical reaction and complete the etching in a shorter time.

Figure 4-4 shows a spray etcher that was built for use in an industrial arts laboratory. This unit can etch a PC board in about 4 minutes. It has a solid-state timing circuit that automatically turns the etchant pump off after the preset time has passed.

There are two etchant solutions commonly used for small batch or single-board applications. Ferric Chloride is the most widely used. It is safe and fairly inexpensive, but it will stain fingers and clothes and so should be handled with care. Ammonium persulfate is another etchant that is easy to use and provides good results. This solution will create heat during the etching process and so must be used with care.

After etching is completed, the board should be rinsed in cold water for at least one minute to remove any traces of the etchant. The resist can then be removed with fine steel wool.

Fig. 4-4. Low-cost spray etcher built for low-volume PC-board fabrication work. This unit contains two gallons of etchant (photograph by Mary Duncan, Oswego Learning Resources Center.)

Drilling

Holes must be drilled in order to mount the components. A high speed drill is usually used for this purpose. A Dremel® hobby tool works quite well in this application. Small drill presses are available to hold the Dremel tool, making the drilling process very easy to control. Drill sizes from #50 to #60 are commonly used for most holes. A #57 will provide the proper hole size for most components. If there is any doubt about the hole size, measure the wire that is to go through the hole and select a drill slightly larger than the wire diameter.

Soldering

After all of the holes are drilled, the components are mounted. Usually, the body of the part is placed in contact with the PC board and the leads are run through the proper holes. The lead is soldered to the copper at the hole and the excess lead is removed with diagonal cutters. Figure 4-5 shows the proper sequence of events for mounting, soldering, and removing excess lead wire when installing a component on a PC board.

A 30-watt soldering pencil is a good size for most PC board applications. Keep the tip clean by wiping it with a damp sponge. Use only rosin (or resin) as a flux. Most solder used for electronics work has a rosin core in the solder. Never use plumbers' solder or acid-paste flux. If acid flux is used, it will damage the solder joint and will eat away the copper much like the etchant did. If a joint turns "green" as it ages it was probably soldered with an acid flux. The rosin flux cleans the joint so the solder can adhere to the wire and the copper. A 60/40 (60 percent tin, 40 percent lead) solder is used for most PC board soldering. This alloy provides a good conductive path and a strong joint.

BREADBOARDING AND EXPERIMENTATION

The following portion of this chapter will cover the practical aspects of experimenting with digital circuits. Two techniques of connecting components will be discussed (wire-wrapping and etched-circuit board), and plans for a breadboard circuit, a power supply, and a component holder will be given. Thorough preparation of these basic building blocks will make it much easier to perform the experiments and exercises that follow in later chapters. Finally, a list of components, such as capacitors, resistors, and integrated circuits, required for later exercises is given at the end of

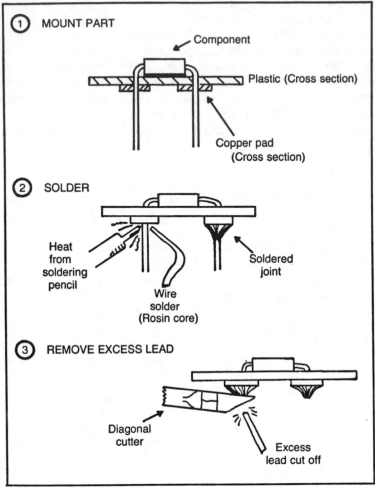

Fig. 4-5. Correct procedure for soldering a component to a PC board.

this chapter. Addresses of several sources where these components can be purchased will be found in the Appendix.

The first step in making a working circuit model is the *breadboard*. This is the test stage, and it consists of connecting all the
component parts together according to the schematic diagram.
Since it is desirable to do this in a temporary fashion, a "breadboard" socket is often used. These sockets are often installed on
units that contain power supplies and other functional devices useful
to the experimenter. Most modern breadboards are perforated
plastic boards designed so that integrated circuits can be "plugged

in" without any difficulty. Figure 4-6 shows a modern breadboard designed especially for digital electronics applications.

Power Supply

A regulated 5-volt power supply is necessary when working with TTL digital ICs. Even if a battery is used, the voltage must be regulated so that voltage variations due to current demands are not felt by the digital-logic circuits.

Figure 4-7 is the schematic diagram of a regulated 5 volt dc power supply that can be used for most TTL experimentation. This power supply, using the 7805 regulator, will deliver a maximum of 750 milliamperes to a circuit. The output is 5-volt dc regulated. If a power supply of 1 ampere output is required, an LM 309K regulator can be used. The transformer and rectifier must also be able to handle 1 ampere. More details on the construction of this power supply are provided later.

Wire Wrap

Wire wrapping is a process that is used extensively in industry (especially in computer applications) and in hobby applications. This process is an alternative to soldering, and it employs a small wire of approximately 36 gauge. Figure 4-8 shows a hand-held unit

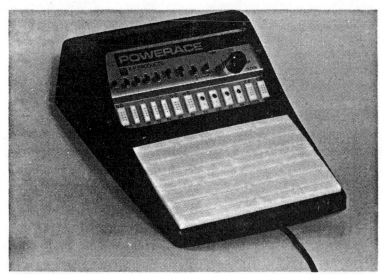

Fig. 4-6. Modern breadboard with power supply and other functions for digital and electronics applications (photograph by Mary Duncan, Oswego Learning Resources Center.)

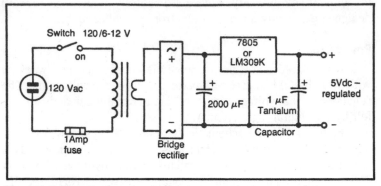

Fig. 4-7. Schematic diagram of a 5-volt dc regulated power supply.

that slits the insulation on the wire so that the wire will make good electrical contact with the junction post around which it will be wrapped. Figure 4-9 shows how the wire is wrapped around a post. The wire is run from terminal to terminal according to a wiring diagram. The terminals are usually numbered so that mistakes are kept to a minimum. Different colors are used for wire insulation, which also helps to prevent mistakes in wiring.

Figure 4-10 shows a more elaborate hand-powered tool that is faster than most other hand units. Power-driven wrappers are useful when a large volume of work is to be done. Wire wrapping requires special pins, boards, and DIP sockets with wire-wrap pins. These are available at most electronics retail stores.

Troubleshooting

When a circuit is to be breadboarded, the wiring should be verified by tracing a red pencil line over the schematic lines. First, each component part is located and plugged into the breadboard socket. Each component is then wired with jumper wires according to the schematic diagram. As each wire jumper is installed, the corresponding line on the schematic should be marked with a red line. When *all* lines on the schematic have been marked, the breadboarded circuit should be complete. The circuit can be tested at this

Fig. 4-8. Slit-n-Wrap tool that slits the insulation of the wire so that it makes good electrical contact with the post on which it is wrapped (courtesy the Vector Electronics Co., Inc.)

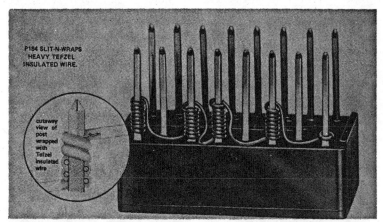

Fig. 4-9. Wire-wrap wiring from terminal to terminal (courtesy the Vector Electronics Co., Inc.)

point. The red-line method may seem tiresome but it will pay off when the circuit works properly the first time.

If the breadboarded circuit does not function properly, the following items should be considered as possible sources of the problem:

1. Check the power supply. Make sure 5 Vdc is available at the

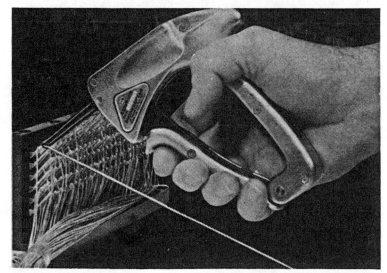

Fig. 4-10. A hand-operated wire-wrap tool (courtesy The OK Machine and Tool Corp.)

proper pins of each IC. Use a dc voltmeter or an LED with a series resistor to check power supply voltages.

2. Be certain that all of the IC pins are inserted into the socket holes. Sometimes a pin is bent under the IC and is not making contact. Remove ICs and check the pins. If any are bent, *carefully* straighten them with small needle-nose pliers and re-insert the IC.

3. Be certain that the ICs are inserted correctly, with pin 1 in the proper socket hole.

4. Check the breadboard socket to see if any holes do not have contacts. In rare instances a contact has been damaged or removed.

5. Check for faulty ICs, resistors, capacitors, etc. Be sure that capacitors are installed correctly if they have polarity markings, especially if large electrolytic capacitors are being used. The best way to test components is to trace the logic signal through the circuit with a digital logic probe or other device that will help trace a digital signal. If such a tool is not available, try to have spare ICs on hand as replacements for questionable components.

6. Re-check wiring with a fresh, unused schematic. Wiring

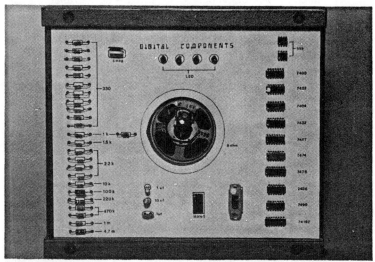

Fig. 4-11. Component storage tray with all components mounted. Each part can be removed and inserted many times before the foam plastic must be replaced (photograph by Mary Duncan, Oswego Learning Resources Center.)

Fig. 4-12. Component trays can be stacked for easy storage. Notice that each tray is numbered for easy identification (photograph by Mary Duncan, Oswego Learning Resources Center.)

errors can easily happen when wiring ICs. It is easy to accidently plug a jumper wire into hole #10 instead of #11. It is a good idea to have another person check your wiring using the red-line process.

Breadboarding is often used to design a new circuit. The plug-in character of this process makes circuit changes easy. Immediate results can be observed, and the circuit can be modified right on the breadboard. When the circuit is complete, a schematic of the operating circuit is drawn so that it can be duplicated at another time.

EXPERIMENTER'S BREADBOARD PROJECT

The following plans for an inexpensive project are provided to aid in performing the exercises in later chapters. This project consists of two parts: the breadboard/power-supply unit, and the component-storage package. Figure 4-11 shows the completed storage tray with components installed. The tray is made of wood with a sheet of ½-inch polystyrene foam cemented to it. The paper showing all parts is rubber-cemented to the plastic foam. When the paper becomes tattered, it can be replaced. The foam can be turned over when the holes become too large from use. Figure 4-12 shows how the trays can be stacked for storage. Figure 4-13 is a layout of the storage insert paper.

The breadboard/power supply is shown in Fig. 4-14. Figure 4-15 shows the interior view with the PC board and transformer visible. Figure 4-16 gives the general dimensions of this unit. A Proto-Board® #100 is used as the breadboard socket. A full-scale layout of both sides of the power-supply PC board is given in Figs. 4-17 and 4-18.

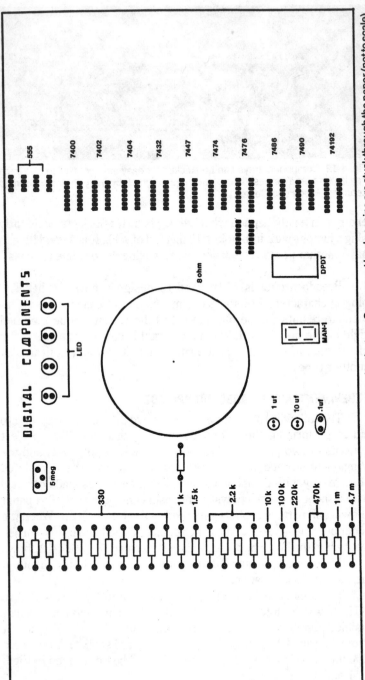

Fig. 4-13. Layout of component storage sheet that can be cemented to the tray. Component leads or pins are stuck through the paper (not to scale).

78

Fig. 4-14. View of completed breadboard/power supply. Notice lead storage compartment on the left side of the unit (photograph by Mary Duncan, Oswego Learning Resources Center.)

Components

The following is a list of parts necessary for completion of the exercises in later chapters. Resistors can be ½-watt, 10 percent tolerance, and capacitors can have a voltage rating of 20 or 25 volts unless otherwise stated.

Fig. 4-15. Interior view of breadboard/power supply showing location of power supply PC board and transformer (photograph by Mary Duncan, Oswego Learning Resources Center.)

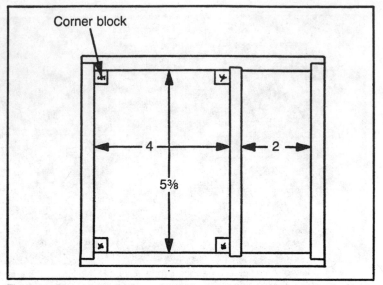

Fig. 4-16. General dimensions of completed breadboard/power supply. Not to scale.

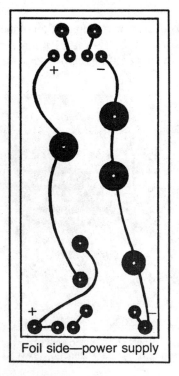

Foil side—power supply

Fig. 4-17. Full-scale layout of foil side of power-supply PC board.

Fig. 4-18. Full-scale layout of component side of power-supply board.

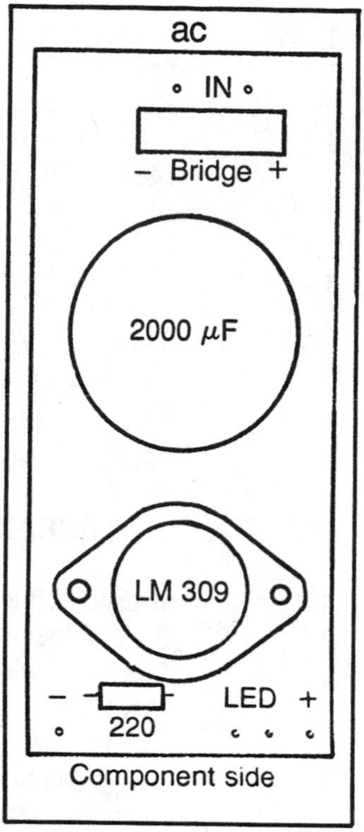

QUANTITY	DESCRIPTION
1	Potentiometer, 5 Megohm
11	Resistor, 330 ohm
2	Resistor, 1000 ohm
1	Resistor, 1500 ohm
4	Resistor, 2200 ohm
1	Resistor, 10-k ohm
1	Resistor, 100-k ohm
1	Resistor, 220-k ohm
2	Resistor, 470-k ohm
1	Resistor, 1 megohm
1	Resistor, 4.7 megohm
4	LED, red
1	Speaker, 2 inch, 8 ohm
1	1 μF Electrolytic capacitor
1	10 μF Electrolytic capacitor
1	.1 μF Disc capacitor

1	MAN-1 Seven-segment LED readout
1	Switch, DPDT slide
2	555 Timer IC
1	7400, Quad 2-input NAND gate IC
1	7402, Quad 2-input NOR gate IC
1	7404, Hex inverter IC
1	7432, Quad 2-input OR gate IC
1	7447, BCD-to-seven-segment driver IC
1	7474, Dual D-type flip-flop IC
2	7476, Dual J-K type flip-flop IC
1	7486, Quad exclusive-OR gate IC
1	7490, Decade counter IC
1	74192, Decade counter, up/down, IC

BREADBOARD / POWER SUPPLY

1	Breadboard socket, Proto-Board #100
1	Fuse holder with 1-ampere fuse
1	Line cord with plug
1	SPST miniature toggle switch
1	Transformer, 120 Vac to 6 or 12 Vac, 1 Ampere
1	Bridge rectifier, 1 ampere, 50 PIV
1	Capacitor, 2000 μF, 25-volt electrolytic
1	LM 309K, 5-volt regulator
1	Resistor, 220 ohm, ½-watt
—	Miscellaneous wood, wire, connectors, PC board stock, solder and hardware

Chapter 5

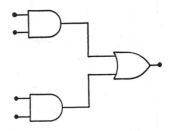

Gate Exercises

Logic gates are basic to all digital operations because they are used to implement the AND, OR, NAND, NOR, and INVERTER functions electrically. The three integrated circuits used in this chapter will allow each function to be proved. An LED will be used as an indicator of the digital output condition of the circuit. When the LED is off, it represents a low (logic 0) condition, and when the LED is on it represents a high (logic 1) condition. This is known as the *positive-logic* system, and will be used throughout the exercises in this chapter.

It is important that an LED never be operated without a current-limiting resistor in series with it. A 330-ohm resistor will usually be sufficient when the power supply is no more than 5 volts. If a limiting resistor is not used, excess current will flow through the LED and burn it out.

The diagrams in Fig. 5-1 are the logic diagrams and pin connections of the three integrated circuits used in the exercises to follow. Always refer to these diagrams when wiring these ICs into the circuit. Make sure that the proper pins are connected to the power supply. Notice that on all three ICs, the negative side of the power supply connects to pin 7, and pin 14 goes to the positive side of the supply.

The exercises in this and later chapters are designed to give you practice in the digital concepts studied in the first four chapters of this book. At the end of each series of exercises, you will find a

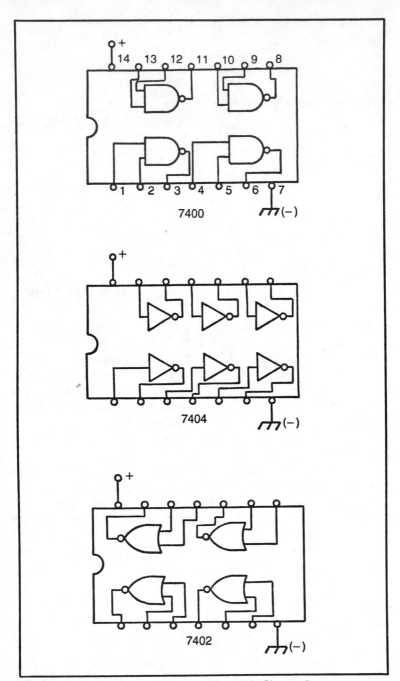

Fig. 5-1. Logic and pin-out diagrams of IC used in Chapter 5.

short quiz designed to test your understanding of the concepts studied. It is as important to complete the quiz and analyze the results as it is to perform the exercise. A poor showing on the quiz means that you should review the theory and do the exercise again. Chapters 2 and 3 cover the theory for the logic functions and circuits used in the exercises. An answer key is provided in the back of this book for a check against your work and the quizzes.

LOGIC GATE EXERCISES

As a result of completing these exercises, you should be able to:

1. Recognize the accepted symbols for digital-logic circuits such as AND, OR, NAND, NOR, and INVERTER.
2. Use common integrated-circuit gates of the TTL family to prove truth tables of common logic operations such as AND, NAND, OR, NOR, and INVERTER.
3. Construct combination gate circuits to implement logic operations of several different types when given the logic function in standard form.
4. Understand, and be able to use, NAND gates to provide OR functions and use NOR gates to provide AND functions

Materials Required

7400 Quad NAND gate (1)
7402 Quad NOR gate (1)
7404 Hex Inverter (1)
LED (light-emitting diode) (1)
330-ohm resistor (1)
Experiment board and power supply (5 volts).

Use a 7400 IC and connect one gate as shown in Fig. 5-2. Make sure that the positive connection of the power supply is connected to pin 14, and the negative connection is hooked to pin 7.

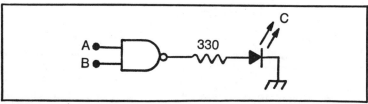

Fig. 5-2.

Use jumpers or leads from the power supply, and connect points A and B as shown in the truth table, Fig. 5-3. Again, positive voltage equals 1 and negative (or zero) voltage equals 0. When the LED is on, it represents a logic at C. If the LED is off, it represents a 0 at C.

Fig. 5-3.

A	B	C
0	0	
0	1	
1	0	
1	1	

1. Complete the C column in Fig. 5-3, using either a 1 or 0 as indicated by the operation of the gate.

2. Write the correct logic expression for the truth table in Fig. 5-3:

   C = _____ (An inverted condition, NOT A, is written $\overline{A}$.)

3. What is name of this operation? _____.

Add to the previous circuit so that a new circuit is constructed as shown in Fig. 5-4.

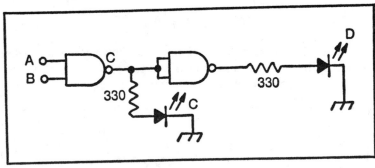

Fig. 5-4.

4. Complete the truth table in Fig. 5-5 by applying logic levels to the inputs and recording the level at the outputs, C and D.

86

A	B	C	D
0	0		
0	1		
1	0		
1	1		

Fig. 5-5.

5. Complete the logic expression for the truth table in Fig. 5-5:

C = _____, and
D = _____.

6. What is the name of the function at D? _____.

Operation

Notice that, although a NAND gate was used, the AND function was the result. In Fig. 5-5, the D function is the AND function. Because the output of the NAND gate was inverted, the N was removed and the result was the AND function. The inverter was developed by wiring the input connections of the NAND gate together. The function demonstrated by this exercise can be drawn as shown in Fig. 5-6. Notice the equality of the three drawings of the same function.

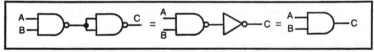

Fig. 5-6.

Wire two gates as shown in Figs. 5-7 and 5-9.

7. Complete the truth tables for each circuit, Figs. 5-8 and 5-10. Make sure the power supply is connected correctly to the IC in use.

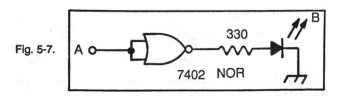

Fig. 5-7.

330

7402 NOR

B

87

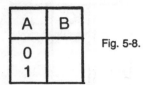

A	B
0	
1	

Fig. 5-8.

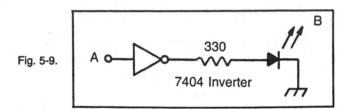

Fig. 5-9.

330

7404 Inverter

B

A	B
0	
1	

Fig. 5-10.

Operation

From the truth table in Fig. 5-8 you can see that the NOR gate also can be used as an inverter. The table (Fig. 5-10) associated with the 7404 inverter shows the same logic as that of the 7402 NOR gate. It is common, when using ICs, to use unused NAND or NOR gates as inverters if the need arises. When many inverters are needed and extra gates are not available, it is best to use an IC designed as an inverter. The 7404 has six such inverters in one package, making it very useful in digital-logic circuits.

Wire the circuit shown in Fig. 5-11 using the 7402 NOR IC.

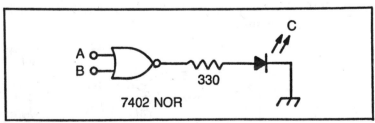

C

330

7402 NOR

Fig. 5-11.

8. Complete Fig. 5-12 by applying logic levels to the gate inputs and recording the output.

Fig. 5-12.

A	B	C
0	0	
0	1	
1	0	
1	1	

9. Write the logic expression for the truth table in Fig. 5-12:

C = _____ or $\overline{C}$ = _____ .

10. What is the name of the logic function shown by Fig. 5-12?

_____ .

Wire the circuit shown in Fig. 5-13.

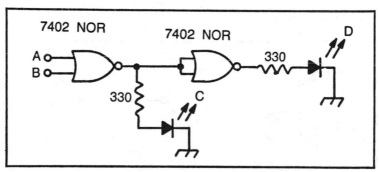

Fig. 5-13.

Apply logic to this circuit and complete the truth table in Fig. 5-14.

Fig. 5-14.

A	B	C	D
0	0		
0	1		
1	0		
1	1		

12. Complete the logic expression for output D in Fig. 5-14:

$$D = \underline{\hspace{3cm}} .$$

13. What is the function of D in Figs. 5-14?

Operation

From the exercise just completed, you can see that by inverting the output of a NOR gate, its function is changed to that of an OR gate. In this case, a NOR gate is used as an inverter by tying both inputs together. This characteristic is the same as was demonstrated by the NAND gate exercise shown in Fig. 5-5. The complete logic function in this case can be redrawn as three equivalent symbolic drawings as shown in Fig. 5-15.

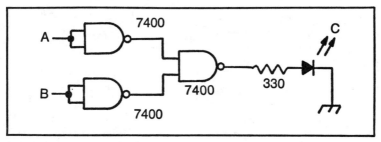

Fig. 5-15.

Wire the circuit in Fig. 5-16.

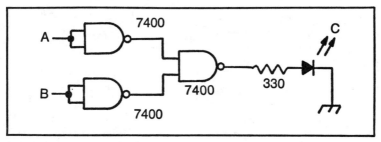

Fig. 5-16.

14. Complete the truth table in Fig. 5-17.

A	B	C
0	0	
0	1	
1	0	
1	1	

Fig. 5-17.

90

15. What is the logic expression for Fig. 5-17? C = _____.

16. What is the function of C? _____.

Wire the circuit in Fig. 5-18.

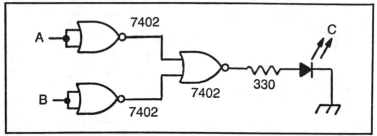

Fig. 5-18.

17. Complete the truth table in Fig. 5-19.

Fig. 5-19.

A	B	C
0	0	
0	1	
1	0	
1	1	

18. What the logic expression for Fig. 5-19: C = _____.
19. What is the function of C? _____.

Operation

You can see from Figs. 5-17 and 5-19 that an AND and OR function can be developed by inverting the inputs to certain gates.

Figure 5-20 shows how an OR function can be obtained from NAND gates.

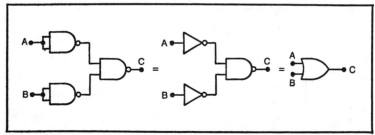

Fig. 5-20.

91

By inverting the inputs to a NAND gate, an OR function results. Figure 5-21 shows how an AND function can be obtained using NOR gates. By inverting the inputs to a *NOR* gate, an *AND* function results.

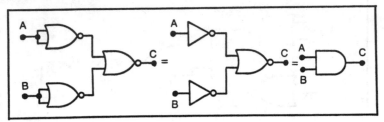

Fig. 5-21.

The procedure above is often used to develop AND and OR gates for digital logic applications. NAND and NOR gates are far more common and useful than AND and OR gates, so the procedure shown here is a very useful one. It allows the development of a gate even if the right one is not available. In any case, the resulting function is what is important.

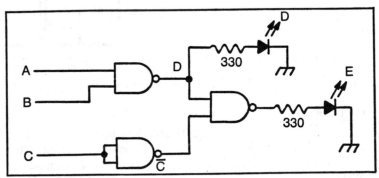

Fig. 5-22.

A	B	C	D	E
0	0	0		
0	0	1		
0	1	0		
0	1	1		
1	0	0		
1	0	1		
1	1	0		
1	1	1		

Fig. 5-23.

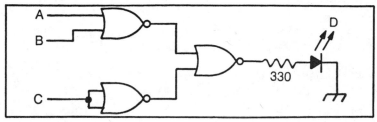

Fig. 5-24.

Wire the circuits given in Figs. 5-22 and 5-24.

20. Complete the truth tables in Figs. 5-23 and 5-25.

Fig. 5-25.

A	B	C	D
0	0	0	
0	0	1	
0	1	0	
0	1	1	
1	0	0	
1	0	1	
1	1	0	
1	1	1	

21. Complete the logic expression for Fig. 5-23: D = _____ and E = _____.

22. Complete the logic expression for Fig. 5-25:

D = _____ .

23. Draw the appropriate symbolic logic circuit for each expression. Use NAND, NOR, or INVERTERS given below.

A. $\overline{A} + BC = D$

B. $A\overline{B} + \overline{C}D = E$

C. $\overline{AB} + C = D$

LOGIC-GATE QUIZ

Select the *one most correct* answer from among those listed. Circle the letter in front of your selection.

1. The symbol used to represent an AND gate function is:

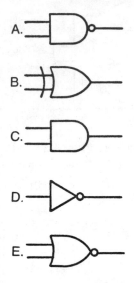

2. Which answer represents the correct Boolean expression for this diagram and truth table?

A. A•B = C
B. $\overline{A}$ + B = $\overline{C}$
C. $\overline{A}$ + $\overline{B}$ = C
D. $\overline{A}$•B = C
E. A + B = C

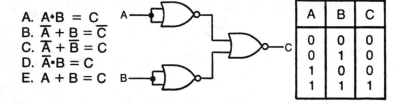

A	B	C
0	0	0
0	1	0
1	0	0
1	1	1

3. Which of the following is *not* an inverter?

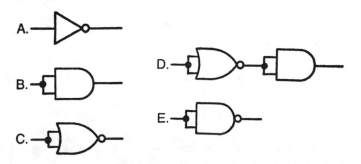

4. Inverting the inputs to a NAND gate changes its function to that of an:

    A. AND gate.
    B. OR gate.
    C. Inverted AND gate.
    D. NOR gate.
    E. Exclusive OR gate.

5. Inverting the inputs to a NOR gate changes its function to that of an:

    A. AND gate.
    B. OR gate.
    C. NOR gate.
    D. Inverted AND gate.
    E. Exclusive OR gate.

6. Which expression listed below is a correct representation of the following statement: NOT A OR NOT B AND C EQUAL D.

    A. $AB + C = D$
    B. $\overline{A} + B + C = D$
    C. $ABC = D$
    D. $\overline{A} + BC = D$
    E. $\overline{A} = \overline{B}C = D$

7. . Which of the following is the correct symbol for a NOR gate?

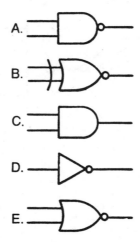

95

8. Draw the symbolic diagram of the following logic statement. Use only NOR or NAND gates and label all inputs and outputs.

$$\overline{A} + BC = D$$

9. Complete the truth table for the following logic gate:

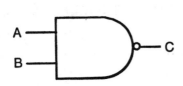

A	B	C
0	0	
0	1	
1	0	
1	1	

10. Complete the truth table for the following logic gate.

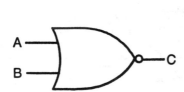

A	B	C
0	0	
0	1	
1	0	
1	1	

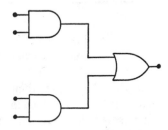

Chapter 6

The Flip-Flop

The flip-flop is a basic logic device with many uses in digital circuits. One use is in counting digitally. Another is in the control of pulse trains—stop or start and debouncing mechanical switches. Flip-flops are also used to monitor digital outputs, and serve as latches, or hold the last digital signal that was sensed. The basic flip-flop is the RS, or reset-set, type, The diagram in Fig. 6-1 shows the RS flip-flop symbolically.

This flip-flop uses two NAND gates, 1 and 2. Suppose the system currently has the output of #2 gate ($\overline{Q}$) is low (0). Since this low is attached to one input of gate #1, the output of gate #1 (Q) must be high (1) if the other input to gate #1 (S) is also low. If the input to gate #2 (R) is brought low, then gate #2 output will go high, driving the gate #1 output to a low condition. In other words, the two outputs must be complements (opposites in logic level) of each other. Both cannot be at the same logic level at the same time. This circuit is monostable, and will remain in the condition it is in unless forced to flip or flop due to a change in the logic signals applied to the R or S inputs.

A disadvantage of this circuit is that if S and R (set and reset or clear) are both given the same logic, the output cannot be controlled. A D-type flip-flop has advantages over the RS type since it can be "clocked" to change the output in response to a pulse from a clock or oscillator. The 7474 IC is a dual-D leading-edge-triggered flip-flop. The circuit and pin layout for this IC are shown in Fig. 6-2.

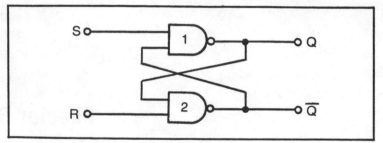

Fig. 6-1.

　　When this type of flip-flop is clocked, whatever logic is on input D is transferred to Q when the clock signal goes from LOW to HIGH (Ground to Positive). If D is high when clocked, Q goes high and $\overline{Q}$ goes low. If D is low when clocked, Q goes low and $\overline{Q}$ goes high. The clear or R (reset), when low, will cause Q to go low and $\overline{Q}$ to go high. If S (set) is low, Q goes high and $\overline{Q}$ goes low. These controls are useful when it is necessary to set or clear a number of flip-flops in a string. R and S should not both be low at the same time or control will be lost. The D type flip-flop is triggered by the edge of a pulse (in this case, the leading edge) so it is called an edge-triggered device.

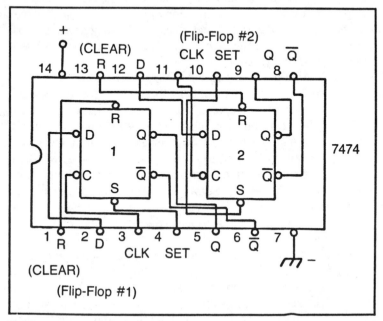

Fig. 6-2.

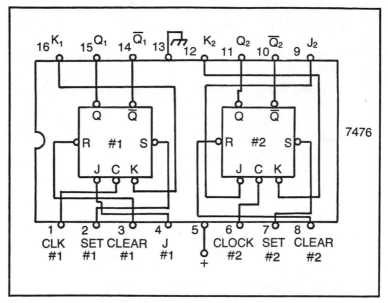

Fig. 6-3.

Another type of clocked flip-flop is the JK flip-flop. The circuit and pin layout of the 7476 dual JK flip-flop are shown in Fig. 6-3.

In this flip-flop, J and K determine conditions under which the clock will cause Q and $\overline{Q}$ to change. If J and K are both low, the clock will have no effect on Q and $\overline{Q}$. They will remain in the logic state they were in before the clock pulse was applied. This is an easy way to hold the output of this flip-flop in a given condition. If both J and K are high, the output will flip-flop, or *divide*, the clock frequency by two.

If J=1 and K=0, the clock pulse will cause Q=1 and $\overline{Q}$=0. If J=0 and K=1, then the clock pulse will cause Q=0, and $\overline{Q}$=1. Again, the clear and set terminals, when low, will clear (Q=0, $\overline{Q}$=1) or set (Q=1, $\overline{Q}$=0). Set and clear should not be low at the same time. Both the 7474 and the 7476 have operating frequencies as high as 20 megahertz.

FLIP-FLOP EXERCISES

As a result of completing the exercises in this chapter, you should be able to:

1. Understand the operational characteristics of both D and JK type flip-flop integrated circuits.

2. Wire and verify actual operation of various flip-flop applications using fixed conditions or clocked logic.
3. Wire and verify the use of an RS flip-flop as a bounceless switch.
4. Wire a JK flip-flop in a simple counter or divide-by-2 operation.

Materials Required

7400 QUAD NAND Gate IC (1)
7474 Dual D Flip-flop IC (1)
7476 Dual JK Flip-flop IC (1)
LED (4)
555 Timer IC (1)
1-megohm resistor (1)
470-K ohm resistor (1)
1-μF capacitor (1)
breadboard and 5-volt dc power supply

The 7400

Use a 7400 QUAD NAND gate IC and wire the circuit shown in Fig. 6-4. Be sure to connect the + and − to the power supply. See the 7400 pin diagram for the correct connections.

1. Ground (−) point A. What is the condition of the LEDs?

$$\frac{Q}{\overline{Q}} = \underline{\hspace{2cm}}$$

2. Remove ground from A and apply it to B. Record LED condition:

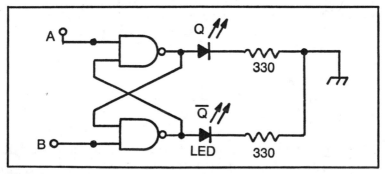

Fig. 6-4.

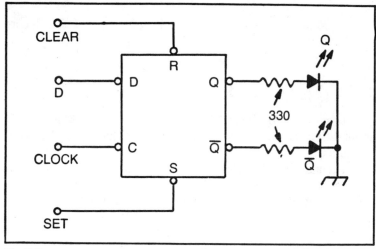

Fig. 6-5.

$$Q = \underline{\hspace{2cm}}$$
$$\overline{Q} = \underline{\hspace{2cm}}$$

Operation

Remove the LED from the $\overline{Q}$ output. By observing the LED at Q, the condition of the flip-flop can be determined since the two outputs are complements of each other. If input A is grounded (say with a push button switch) Q should be "on" or high, and if B is momentarily grounded (low), Q will be "off" or low. This circuit can be used as a bounceless switch because the switching from high to low or low to high is very fast. Mechanical switches "bounce," often causing undesirable effects in digital electronics. This circuit is often used to avoid the bounce of mechanical switches and to condition (clean up) digital signals.

The 7474

Wire the 7474 IC (one flip-flop) according to the diagram shown in Fig. 6-5. Refer to the flip-flop pin diagram (Fig. 6-2) for the proper pin connections. Be sure to attach the power leads.

3. Ground (low) the CLEAR Input. Record the LED condition:

$$Q = \underline{\hspace{2cm}}$$
$$\overline{Q} = \underline{\hspace{2cm}}$$

4. Remove ground from CLEAR and ground SET. Record LED condition:

$$Q = \rule{2cm}{0.4pt}$$
$$\overline{Q} = \rule{2cm}{0.4pt}$$

Leave CLEAR and SET open and ground D. Momentarily ground SET so that $Q = 1$ and $\overline{Q} = 0$. (This is the "Before" condition.)

Ground CLOCK. When this ground is broken, the clock logic will rise to a high condition. This causes a low-to-high transition of the clocking. A bounceless switch can also be used for the clock application.

5. Record LED condition:

Before	After
$Q = 1$	$Q = \rule{2cm}{0.4pt}$
$\overline{Q} = 0$	$\overline{Q} = \rule{2cm}{0.4pt}$

Now make D high (1), and momentarily ground CLEAR. This sets up the "Before" condition.

6. Momentarily bring CLOCK high (1) and record LED condition:

Before	After
$Q = 0$	$Q = \rule{2cm}{0.4pt}$
$\overline{Q} = 1$	$\overline{Q} = \rule{2cm}{0.4pt}$

Operation

The preceding exercise should verify that grounding the SET input will bring the flip-flop output to $Q=1$ $\overline{Q}=0$, and that grounding the CLEAR terminal will cause the output to go to a $Q=0$ $\overline{Q}=1$ logic condition. It also shows that the level on D will be passed on to the Q output when the clock signal input goes from low to high. This is known as leading-edge triggering of the device.

The 555

Wire the 555 timer and the 7474 according to the diagram in Fig. 6-6. The 555 "clock" will deliver a square wave signal at about 1 hertz to the 7474.

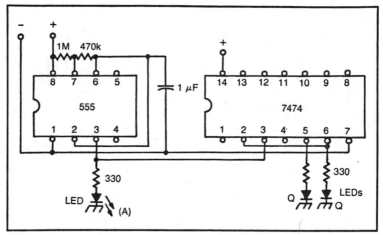

Fig. 6-6.

Turn the power on and observe the three LEDs.

7. Count 10 pulses of A and fill in the number of pulses of Q and $\overline{Q}$ in Fig. 6-7.

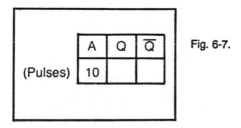

Fig. 6-7.

	A	Q	$\overline{Q}$
(Pulses)	10		

8. Does the A pulse count have the same frequency as Q and Q ? _____ .

Operation

This circuit can be used to turn on and off automatically in response to the clock. If several clock pulses arrive at the same time, it can be made sensitive to any combination that will provide a positive leading edge. Since the output "follows" the input or clock, this circuit momentarily watches the input. If the high level were removed from the D, the output would remain at the logic condition of the last pulse that caused a flip-flop. This characteristic is called latching, and it is useful as a "memory" device. To further explore this "pass on" effect, complete the following exercise.

103

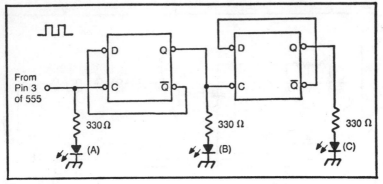

Fig. 6-8.

Cascaded Flip-Flops

Wire the 7474 as shown in Fig. 6-8. Simply move the 330-ohm resistor and the LED to pin 9 and jumper from pin 5 to 11. Tie pin 12 to pin 8 and pin 2 to 6. This circuit is now one flip-flop driving a second flip-flop. See the IC pin diagrams for the correct pin connections.

Stop the clock after each pulse so that the data can be observed. Stop and start the clock (555) by removing the wire to pin 6. Set the system for A, B and C all equal to 1 (high). Pulse the input and record the logic state changes as they occur.

9. Record the logic changes in Fig. 6-9.

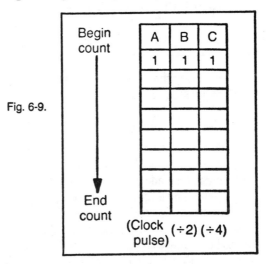

Fig. 6-9.

10. Does A pass on to B ? _____ .

11. Does B pass on to C ? _____ .

Observe the data recorded in Fig. 6-9.

12. What relationship exists between the frequency of A and C ? (A = _____) (C = _____) .

Operation

From the operation of this circuit and the data collected, it should be apparent that the signal at A is passed on to B (and the signal at B is passed on to C). This means that the flip-flops are in series and are passing on the information at D (high level) on each positive pulse from the clock. The circuit can be pulsed differently (two clocks, one for each flip-flop) providing an independent output from each flip-flop. While the circuit is in operation, remove the D connection to each flip-flop alternately. In each case the other flip-flop will continue to follow the input and the one whose D has been removed from the high level will hold the last logic condition present before the high level was removed. This is the memory application (latching) mentioned earlier. Usually, a latch uses four flip-flops in one package in order to follow and hold, whenever instructed, a 4-line or BCD-type parallel signal such as would be used to drive a decimal readout. This allows a digital counter to keep counting while the latch holds the last display on the readout device. A digital stopwatch is a typical application of this principle.

The 7476

Wire the 7476 JK flip-flop IC according to the diagram in Fig. 6-10. Use the previously set up 555 timer for a 1 hertz clock. Notice that the power supply pin connections on the 7476 are at different locations than on previous ICs.

Turn the power on and observe the LED.

13. Record the information about the frequency of the circuit in Fig. 6-11. Count pulses at B for each 10 pulses at A.

14. What is the relationship between A and B?

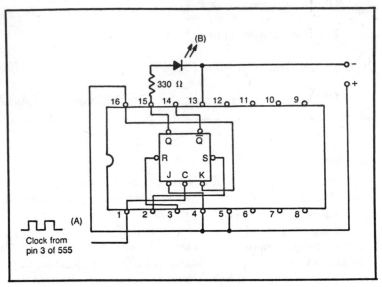

Fig. 6-10.

$$A = \underline{\hspace{2cm}} B$$

15. What function is being performed from A to B in Fig. 6-11
_____ .

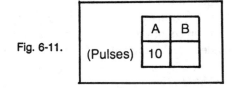

Fig. 6-11.

(Pulses)	A	B
	10	

Operation

By changing J and K, the flop output can be inverted. If J is high and K low, clocking makes Q high and $\overline{Q}$ low. If J is low and K is high, clocking will make Q low and $\overline{Q}$ high. This can be verified by experimenting with the JK flip-flop IC used in this exercise. The set and reset (clear) will each operate when grounded or brought to a low logic condition. They should not be grounded at the same time. The circuit in Fig. 6-10 is actually dividing by two because the clock output at B pulses at half the frequency of the input clock signal. This flip-flop is very effective as a counter. It can be wired to more complex internal circuitry to provide counters and dividers that

106

operate on numbers greater than two. The most useful application of the JK flip-flop is in data control. The J and K terminals give it the versatility and adaptability necessary for data transfer. Since there are two JK flip-flops in the 7476 pack, a divide-by-four device can be constructed if one flip-flop is used to furnish the input to the second.

FLIP-FLOP QUIZ

Select the one *most correct* answer from among those listed. Circle the letter in front of your selection.

1. The outputs (Q and $\overline{Q}$) of a common flip-flop are:

  A. Supplements.
  B. Tandem.
  C. Asymmetrical.
  D. Complements.
  E. Astable.

2. The JK flip-flop responds to the

  A. Clock pulse alone.
  B. Clock pulse and condition of J and K.
  C. Level of D.
  D. J level only.
  E. Clock pulse duration.

3. Which of the following flip-flops is generally used as a latch?

  A. JK type.
  B. RS type.
  C. FF type
  D. CP type.
  E. D type.

4. The flip-flop is used as a basic unit in digital

  A. Registery.
  B. Clocking.
  C. Graphing.
  D. Sensing.
  E. Counting.

5. In order to divide by 10, how many flip-flops would need to be connected serially? (select the smallest number that will work)

A. 2 (divide by 5 and divide by 2).
B. 5 (each divide by 2).
C. 10 (divide by 5 each).
D. 20 (divide by two).
E. 50 (divide by 5 types).

6. In the circuit below, if a negative pulse were applied to A, Q and $\overline{Q}$ would be:

  A. $Q = 0, \overline{Q} = 1$.
  B. $Q = 0, Q = 0$.
  C. $Q = 1, \overline{Q} = 1$.
  D. $Q = 1, \overline{Q} = 0$.
  E. Undetermined.

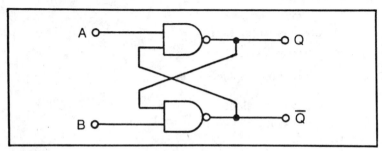

Fig. 6-12.

7. Using the diagram from question 6, if B were made negative, Q and $\overline{Q}$ would go to:

  A. $Q = 1, \overline{Q} = 0$.
  B. $Q = 0, \overline{Q} = 1$.
  C. $Q = 1, \overline{Q} = 1$.
  D. $Q = 0, \overline{Q} = 0$.
  E. Undetermined.

8. If 10 flip-flops are wired serially and you wish to CLEAR all to a $Q = 0, Q = 1$ condition, which of the following would be the best method to use?

  A. Use the clock to pulse all to $\overline{Q} = 0, \overline{Q} = 1$.
  B. Apply SET pulse to all simultaneously.

C. Turn power off momentarily.
D. Increase supply voltage until clear condition results.
E. Apply RESET pulse to all simultaneously.

9. In the following JK flip-flop, J and K are both low (0). Which waveform represents the correct operation of Q?
(Q = 0, Q = 1 before clock pulse is applied)

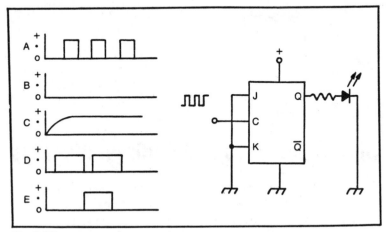

Fig. 6-13.

10. If both J and K of the JK flip-flop in question 9 are tied to positive (+) or high, the output of Q will be:

   A. Division of clock by 2.
   B. High (1).
   C. Low (0).
   D. Double the input frequency.
   E. Duplicate of the input frequency.

Chapter 7

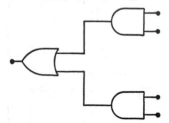

Digital Counters and Decimal Displays

Counting is an important part of digital devices such as computers, frequency counters, and other instruments. Basically, counters are of two types: synchronous (which will count upon receipt of a clocked pulse) and asynchronous (which respond immediately to changes in input.) Asynchronous is sometimes called a direct system. This chapter is concerned with the direct counter. Clocked or synchronous counters are useful in systems where updating or sampling, such as in frequency counters, is important.

Since digital counting is done using binary codes (the BCD code in particular), and since the decimal system is commonly used as a display, it is necessary to convert between the two. This conversion is usually done by a decoder with a seven-segment output. The seven-segment output directly drives a display device such as an LED readout.

COUNTER EXERCISES

As a result of completing the exercises in this chapter, you should be able to:

1. Use the 7490 decade counter integrated circuit.
2. Understand binary counting and the BCD code.
3. Understand BCD-to-decimal decoding using the 7447 BCD-to-seven-segment decoder driver IC.
4. Use a presettable up/down counter such as the 74192.

Materials Required

> 7490 decade counter IC (1)
> 7447 BCD-to-seven-segment decoder driver IC (1)
> 330-ohm resistor (11)
> MAN-1 common anode seven-segment LED readout (1)
> 74192 presettable up/down counter IC (1)
> 7400 Quad NAND gate IC (1)
> SPDT toggle or push button switch (1)
> 2.2-k ohm resistor (2)
> Breadboard and 5-volt power supply.
> LED (4)

The 7490

Wire tne 7490 IC as a decade (count-to-10) counter by following the diagram in Fig. 7-1. Wire the 7400 NAND Gate as a bounceless switch for pulsing the counter. The LEDs will be used to read the output (BCD) states of the counter at each count.

When the circuit in Fig. 7-13 is turned on, the LEDs may light in any order. All LEDs should be turned off (0) by breaking the ground connection momentarily at pins 2 and 3 of the 7490 IC. Set the bounceless switch to a "0" position.

1. Pulse the counter (turn switch to 1 and 0 alternately) and record the LED output states in Fig. 7-2. If an LED is *on*, record a 1. If an LED is *off*, record a 0.

2. Explain what happens to the binary output after number 9 pulse has been passed.

Operation

This exercise should have verified that the 7490, in this case, counts to ten in a BCD code, and then recycles. The BCD output can be used in its present form or it may be decoded for other digital applications. If a single pulse is needed at the end of the decade or ten count, the 7490 can be wired in a different way. The BCD code at ten can be decoded using gates to develop a single pulse at the end of a decade count.

The 7447 and MAN-1

Remove the LEDs from the 7490 and wire the 7447 and MAN-1 readout according to the diagram shown in Fig. 7-3. Be careful to

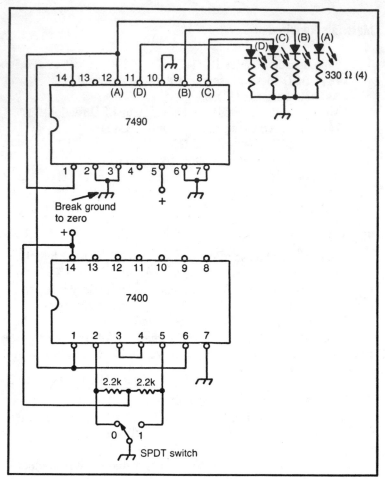

Fig. 7-1. Wiring diagram for a BCD-output decade counter.

avoid shorts between the 330-ohm resistors. Connect the BCD output from the 7490 to the input of the 7447.

After the circuit in Fig. 7-3 is wired, zero the 7490 output by momentarily removing the ground from pins 2 and 3. The readout should now read "0."

Pulse the 7490 with the switch as in the previous step and observe the MAN-1 readout after each pulse. Pulse the counter through several cycles so that its operation is clearly understood.

Operation

The BCD output from the 7490 is converted by the 7447 into a seven-segment code that drives the MAN-1 segments. The segments combine to produce the decimal numbers 0 to 9. This type of system can count very rapidly and is very useful, but it will not display the alphabet. The addition of a memory, or latch, between the 7490 and the 7447 can be used to hold a count on the display while the 7490 continues to count, making this circuit useful in applications such as stopwatches and multiple-event counters.

The 74192

Replace the 7490 with the 74192 presettable up/down counter. Notice that the wiring is different and that the 74192 has 16 pins, whereas the 7490 has 14. Carefully rewire the 74192 circuit according to the diagram shown in Fig. 7-4.

Use the code shown in Fig. 7-5 to set each code on the input pre-load terminals of the 74192 (L_1, L_2, L_4, L_8). Momentarily bring the load (pin 11) to ground after coding the pre-set data in each case. After a code is entered, pulse the counter and return the counter to any number other than the next one to be coded into the load.

3. Complete the decimal column in Fig. 7-5.

Remove the + from pin 4 of the 74192 and attach it to pin 5. (Exchange pin 4 and pin 5 wiring so that now pin 4 = input from 7400 and pin 5 = +). Proceed to load several BCD numbers into the pre-load of the 74192 and pulse the circuit as in the previous step.

Fig. 7-2.

Pulse No.	LEDs			
	D	C	B	A
0				
1				
2				
3				
4				
5				
6				
7				
8				
9				

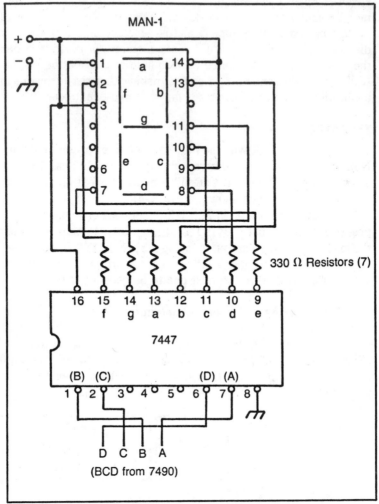

Fig. 7-3. Wiring of 7447 decoder/driver and MAN-1 LED readout.

4. Does the count proceed in an up pattern? _____.

5. In what way does the count proceed in this case ?
_____.

Operation

The 74192 has some very desirable features. It can be programmed or loaded with pre-set numbers in BCD code. This can be done manually, as in this exercise, or with BCD thumb-wheel switches. This counter can also count either up or down so it can be

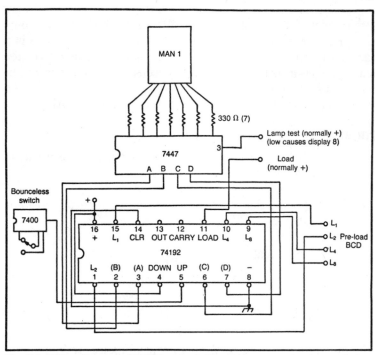

Fig. 7-4. Wiring diagram of 74192 presettable up/down counter demonstrator.

used in applications where frequent "sampling" of frequencies going both ways is necessary. If the frequency were moving up and down, this unit could record the count at any given time by comparing the

PRESET (BCD)				DECIMAL
L_8	L_4	L_2	L_1	
0	0	0	0	
0	0	0	1	
0	0	1	0	
0	0	1	1	
0	1	0	0	
0	1	0	1	
0	1	1	0	
0	1	1	1	
1	0	0	0	
1	0	0	1	

Fig. 7-5.

frequency with the previous count. A considerable amount of outside circuitry would be necessary in order to do this. Pin 12 is a carry connection used to *cascade* (connect serially) another counter unit. Pin 13 is a borrow output that is used for the same purpose in the down-count mode. Pin 14, when high (1), will clear the counter to "0." This pin must be grounded during normal countring.

COUNTER QUIZ

Select the one *most correct* answer from among these listed. Circle the letter in front of this selection.

1. A binary code that is commonly used when counting to ten digitally is:

    A. Hexadecimal.
    B. Grey.
    C. BCD.
    D. Excess three.
    E. Octadecimal.

2. When using a TTL counter, the signal (pulse) train must be "conditioned" such as with a bounceless switch. This conditioning is used to:

    A. Keep the voltage stable.
    B. Prevent counting error pulses when turning on or off.
    C. Reset the counter flip-flops.
    D. Prevent accidental turn off of train.
    E. Prevent heat build-up when the chip remains on.

3. The numbers 10 to 15 are indicated as (these are the invalid numbers in the BCD code, but when decoded and displayed in a seven-segment-format readout are represented by . . . . .):

    A. 10, 11, 12, 13, 14, 15.
    B. a, b, c, d, e, f.
    C. 01, 001, 010, 011, 100, 110.
    D. 01, 01, 03, 04, 05, 06.
    E. A_1, B_1, C_1, D_1, E_1, F_1.

4. The purpose of the 7447 TTL integrated circuit is to drive an LED readout and also to:

A. Provide a load resistance for the LED readouts.
B. Act as a latch or memory for the counter.
C. Buffer or store data in parallel form.
D. Isolate the pulse train from the LED readout.
E. Convert the BCD code into a seven-segment code.

5. The diagram below is of an LED readout using seven LEDs. This is a configuration known as a common:

A. Cathode.
B. Base.
C. Input.
D. Anode.
E. Current.

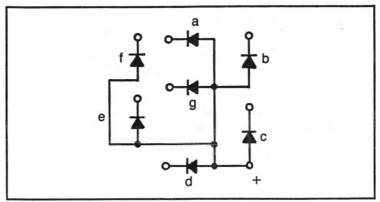

Fig. 7-6.

6. The 7490 counter is basically a (an):

A. Down counter.
B. Up-down counter.
C. Up counter.
D. Reset counter.
E. Set counter.

7. Why must resistance be placed in series with the LED segments of a seven-segment LED readout?

A. To limit LED brightness.
B. To bias the LED.

117

C. To provide a load for the driver.
D. To increase current to the LED.
E. To limit the LED current and prevent burnout.

8. The 7490 consists of two counters that may be used indepen-
dently or cascaded. One is a divide-by-2 clock and the other is a
divide-by-:

A. 3.
B. 4.
C. 5.
D. 8.
E. 10.

9. The "0 SET" pins of a 7490 are used to:

A. Test the BCD output.
B. Set the counter to 8.
C. Test the decimal point.
D. Eliminate leading zeros in a long series.
E. Set the counter to zero.

10. Suppose a circuit uses several dividing counters in a manner
such that one counts by 5, and this output drives a divide-by-2
counter so that the result is a divide-by-10 system. This is
called a:

A. Coded counter.
B. Ripple counter.
C. Down counter.
D. Simultaneous counter.
E. Parallel counter.

Chapter 8

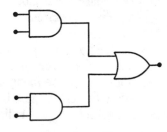

Shift Registers

The shift register is another application of the flip-flop, usually JK or D type. Shift registers are useful for converting serial data to parallel form or parallel data to serial form. The following exercises explore JK flip-flop applications, since it is the most useful for digital counting.

SHIFT-REGISTER EXERCISES

As a result of completing the exercises in this chapter, you should be able to:

1. Understand the use of the JK flip-flop in a shift register application.
2. Use the shift register to convert data from parallel to serial or serial to parallel form.
3. Understand the principle and operation of a ring counter.

Materials Required

LED (4)
330-ohm resistor (4)
7476 dual JK flip-flop IC (2)
555 timer IC (1)
1-μF capacitor (1)
220-k ohm resistor (1)

4.7-k ohm resistor (1)
470-k ohm resistor (1)
5-volt power supply and breadboard.

4-Bit Shift Register

Wire the circuit shown in Fig. 8-1. The 555 IC is wired as an astable square-wave clock. Use the slow (470-k ohm resistor) speed to pulse the circuit at first, and the fast speed later to see the ripple effect of the ring counter. Do not wire the "dashed" lines at this time. These wires will be installed later for the ring counter application. This circuit is a 4-bit shift register. Only the Q output of each flip-flop will be monitored with an LED.

After the circuit is wired, turn the power on and clear the register to 0000 by grounding the CLEAR terminal.

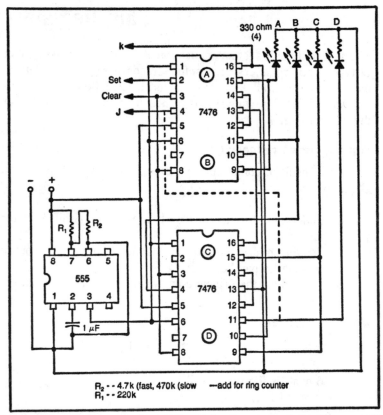

Fig. 8-1. Wiring diagram for 4-bit shift register. Do not wire dashed lines except when operating as a ring counter.

Clock	Flip-flop state			
pulse	A	B	C	D
Start 0	1	0	0	0
1				
2				
3				
4				

Fig. 8-2. Truth table for circuit shown in Fig. 8-1.

Make J=0 and K=1. This will allow the first clock pulse to cause the "A" flip-flop to change states.

Enter a 1 into flip-flop "A" by momentarily grounding the SET terminal. The "A" LED should now light, and the register should read 1000.

To disable the 555 clock, remove the connection from pin 6 to pin 7 at either end. When the clock is to be operated, connect this wire again.

Activate the clock momentarily so that only one pulse is delivered to the register. After each pulse, stop the clock and record the logic state of the register in Fig. 8-2. In order to re-enter data into flip-flop A, ground the SET terminal. Figure 8-3 is a simplified block diagram of this shift register.

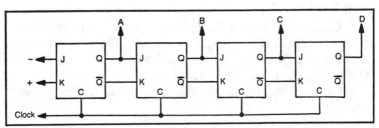

Fig. 8-3. Block diagram simplification of 4-bit shift register.

1. Describe the action of the LEDs according to the data in Fig. 8-2.

2. What happened at each clock pulse?

3. How many pulses were necessary in order to return the register to zero (0000)? _____.

Operation

This exercise demonstrated the "pass on" or shifting function of this circuit. Each flip-flop enables the next flip-flop in line, and when clocked, passes on the 1. Remember that the J must be a 1 and K a 0 at the time of clocking in order for the flip-flop to change state to Q=1, $\overline{Q}$=0, and that if J=0 and K=1, then the clock pulse will cause Q=0 and $\overline{Q}$=1. All CLEAR connections are wired together so that all flip-flops can be cleared to zero at once. This clearing also represents entering data in parallel form. If desired, data could be entered as a 1 or 0 into each flip-flop by properly "setting" each flip-flop. Once data is entered it represents a parallel "word" since it can be viewed or read all at once. If the LSB (least significant bit) flip-flop (D) data is monitored as the clock is pulsed, the data shifts to the right from flip-flop to flip-flop until the register is cleared to 0000. This represents a serial output (bit-by-bit) of the data that was entered in parallel form, making this circuit a parallel-to-serial converter. In order to enter data in serial form, it is shifted into flip-flop "A" bit-by-bit and shifted to the right until the entire word is registered. By observing the Q status of each flip-flop (the LEDs) one can see the parallel form of the data that has been entered, making this circuit a serial-to-parallel converter. The size of the word to be registered determines how many flip-flops are needed, since each flip-flop stores 1 bit of data. Shift registers also serve as memory devices because of their ability to store digital data in either serial or parallel form.

Ring Counter

Change the wiring of the shift register circuit by adding jumper wires from pin 11 of flip-flop D to pin 4 of flip-flop A and from pin 10 flip-flop A, ground the SET terminal. Figure 8-4 is a simplified block diagram of this shift register.

Enter a 1 into flip-flop A, then turn on the clock to pulse the register.

4. Record the logic data for 10 pulses in Fig. 8-4.

5. Explain what occurred during this exercise. Use the data in Fig. 8-4.

6. Did the register eventually clear to 0000?

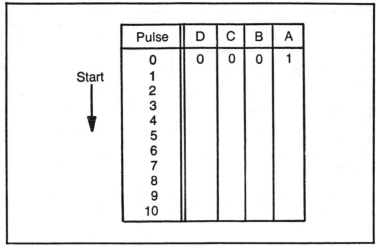

Pulse	D	C	B	A
0	0	0	0	1
1				
2				
3				
4				
5				
6				
7				
8				
9				
10				

Start ↓

Fig. 8-4. Truth table for ring counter.

Operation

You probably noticed that just when the shift to the right was about to clear the register, the "D" flip-flop (which is now hooked to the "A" flip-flop) entered a 1 into flip-flop "A." The shifting began again and would continue on and on. This circuit is known as a ring counter. It forms the basis of many counters used in digital electronics. For example, one could (with these 4 flip-flops) count to 4 (shift four places) and automatically begin over and count to 4 again. The D flip-flop would go to a 1 state for every 4 input pulses provided a 1 has been loaded into the system at A. If flip-flops are added (or other steering gates used) shift registers which will count all sorts of combinations can be constructed. The decade counter is one such flip-flop application that is very useful in digital circuits. This is because the use of the base-10 number system often requires counting by 10's.

If the fast resistor (4.7-k ohm) is used to operate the 555 clock, and the shift register is wired as a ring counter, the result is a fast-moving light effect similar to that of old-fashioned theater marquees. Any number of unusual effects can be arranged with a clock and a shift register.

SHIFT-REGISTER QUIZ

Select the one *most correct* answer from among those listed. Circle the letter in front of the selection.

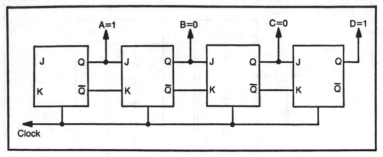
Fig. 8-5.

1. The data indicated above from the 4 flip-flop register at A, B, C, and D, is in what form?

    A. Serial.
    B. Parallel.
    C. Shift.
    D. Complete.
    E. Set.

2. The data to be shifted into the register in question 1 must be shifted into the device bit-by-bit and would be what form?

    A. Serial.
    B. Parallel.
    C. Shift.
    D. Bit.
    E. Set.

3. By setting all the flip-flops of a register at once you are entering data in which form?

    A. Serial.
    B. Parallel.
    C. Set.
    D. Register.
    E. Pre-set.

4. One important use of a shift register is as a

    A. BCD generator.
    B. Multivibrator.

C. Memory.
D. One-shot monostable clock.
E. Grey-code decoder.

5. A type of shift register that takes the last shift data and re-enters it into the entry flip-flop is a type of counter known as a

A. BCD counter.
B. Grey counter.
C. Modulo counter.
D. Ring counter.
E. Decimal counter.

Chapter 9

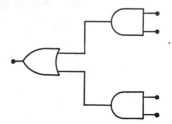

The Exclusive-Or Gate

The Exclusive-OR gate is important in such applications as adders, comparators, code-converters, and parity checkers. This chapter will explore the parity-checking and digital-comparator functions of the Exclusive-OR gate.

EXCLUSIVE-OR GATE EXERCISES

As a result of completing the exercises in this chapter, you should be able to:

1. Understand the specific logic characteristics of the Exclusive-OR gate.
2. Use the Exclusive-OR gate as a data parity checker.
3. Use the Exclusive-OR gate as a digital word comparator.
4. Understand the characteristics of the 7486 Quad Exclusive-OR Gate IC.

Materials Required:

7486 Quad Exclusive-OR gate IC (1)
LED (1)
330-ohm resistor (1)
7432 Quad 2-input OR gate (1)
5-volt power supply and breadboard.

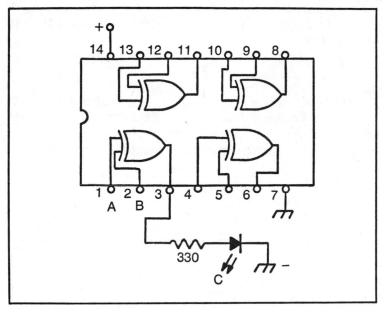

Fig. 9-1. 7486 Exclusive-OR gate.

The 7486

Wire the 7486 according to the diagram in Fig. 9-1.

1. Complete the truth table in Fig. 9-2.

A	B	C
0	0	
0	1	
1	0	
1	1	

Fig. 9-2. Truth table for the 7486.

2. How does the completed truth table (Fig. 9-2) compare with the truth table for an OR gate?

Operation

The Exclusive-OR function, as the truth table indicates, provides a 1 (high) output only when the two inputs are different in logic level. The symbol used to designate the Exclusive-OR function is a circle with a plus sign inside. The function in the truth table is expressed as:

$$A \oplus B = C$$

This function is a very useful and powerful tool in digital electronics. For example, binary numbers can be compared to determine if they are the same or different, or they can be checked to determine if an error is present in a particular group of bits.

Parity Checker

Wire the Exclusive-OR gate as shown in Fig. 9-3. This is a parity checker or detector.

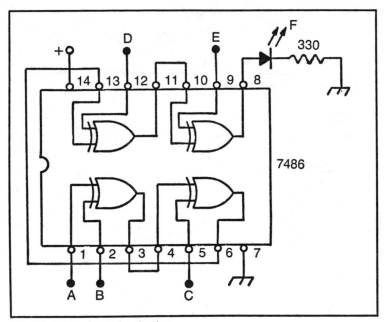

Fig. 9-3. Wiring for parity checker.

3. Complete the truth table in Fig. 9-4.

4. Observe the odd and even number of ones in the ABCDE column and compare this with column F.

5. When an add number of ones appear in ABCDE, F is always _____ .

A	B	C	D	E	F
0	1	1	0	1	
1	1	0	1	1	
1	1	1	0	1	
0	1	1	0	0	
1	0	0	0	0	
1	1	1	1	1	
1	0	0	1	0	
0	0	1	1	1	
1	0	1	0	1	

Fig. 9-4. Truth table for parity checker.

6. When an even number of ones appear in ABCDE, F is always _____.

Operation

It should be obvious that an odd-even detector for a 5-bit binary word is being used in this exercise. This function is known as parity. The circuit can determine whether an odd or even number of ones are contained in the 5-bit word being checked. As a practical application, teleprinters often use such a parity checking system to detect errors. A teleprinter code (such as the 8-bit ASCII) is arranged so that a parity bit is added to each word so that it will have either an even or odd number of ones. As each word is received, it is examined by the parity checker. If the word does not have the correct parity (due to switch bounce, malfunction, etc.) it is detected as an error. Since the data is moving at a very fast rate, knowing that the error exists is of little help, so in some cases the parity checker is used in conjunction with other circuitry to actually correct the error and recreate the word that was changed during transmission.

The 7486 and 7432

Wire the 7486 and 7432 ICs according to the diagram in Fig. 9-5.

7. Complete the truth table in Fig. 9-6.

Compare word A with word B and answer the following:

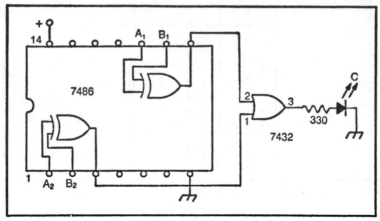

Fig. 9-5. The 7486 and 7432 as a word comparator.

Word A		Word B		
A₁	A₂	B₁	B₂	C
0	0	0	0	
1	0	1	0	
0	1	0	1	
1	1	1	1	
0	0	1	1	
1	0	0	1	
0	1	1	0	

Fig. 9-6. Truth table for the word comparator.

8. When word A = word B, C = _____ .

9. When word A = word B, C = _____ .

Operation

It should be evident from Fig. 9-6 that when words A and B are exactly the same, then the output C is low (0), and when A and B are not the same, the output C is high (1). This is a comparison of digital words. An example of an application of this type circuit is a case where one wants to "look at" a large number of binary words and count the frequency of occurrence of a particular word. This sort of circuit can be used to detect the word each time it appears as the data is being scanned. A counter attached to the output would simply count each time the word appeared.

The Exclusive-OR function is also useful in converting the BCD code into the Gray code, which is a special code developed for electromechanical devices. Since both codes are frequently used in digital circuits, converting from one to the other is necessary. The exclusive OR performs this function accurately, effectively, and with very few components.

This gate might also be used as a combination-number lock since the digital word comparison function is necessary for such an application.

EXCLUSIVE-OR GATE QUIZ

Select the one *most correct* answer from among those listed. Circle the letter in front of your choice.

1. The Boolean symbol that represents the Exclusive-OR function is:

    A. >>.
    B. +.
    C. $\odot$.
    D. $\oplus$.
    E. $\wedge$.

2. In an Exclusive-OR gate, the output will be high (1) only when the inputs are:

    A. The same.
    B. Different.
    C. Low.
    D. High.
    E. Floating.

3. Parity refers to an equality that may be odd or even. In the truth table (Fig. 9-6) which answer indicates an EVEN parity condition at C?

    A. A.
    B. B.
    C. C.
    D. D.

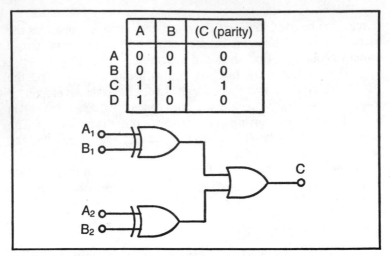

	A	B	(C (parity)
A	0	0	0
B	0	1	0
C	1	1	1
D	1	0	0

Fig. 9-6.

4. Parity is used in digital electronics to:

    A. Detect errors.
    B. Count.
    C. Create codes.
    D. Correct key-bounce.
    E. Disable read-outs.

5. The following circuit represents a:

    A. AND gate.
    B. Parity generator.
    C. Pulse shaper.
    D. Bi-stable switch.
    E. Word comparer.

6. The logic symbol for an Exclusive-OR gate is:

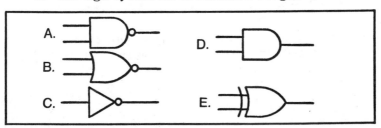

Fig. 9-7.

Chapter 10

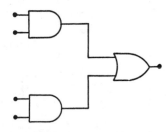

A Digital System

This chapter contains an exercise that combines the 555 timer IC, a 7490 counter, a 7447 decoder/driver, and an LED readout. The result is a practical, working digital system that both provides a useful instrument and illustrates the principles discussed in this book. This system will be a digital-readout ohmmeter that can be calibrated. The block diagram of this simple system is shown in Fig. 10-1.

In operation, the readout is "zeroed" by placing a positive voltage (high) on the proper pin of the 7490 counter. An unknown resistance (R) is applied to the 555 monostable and the monostable is activated by making pin 2 low momentarily. The "on" time of this monostable is dependent on the value of the unknown R. While the monostable is "on," its output provides the positive (high) pulse necessary to activate the astable 555. This second multivibrator will generate a series of pulses. These pulses are counted and displayed by the counter.

THE PROJECT

As a result of completing the project in this chapter, you should be able to:

1. Recognize a system application using a monostable and astable multivibrator.

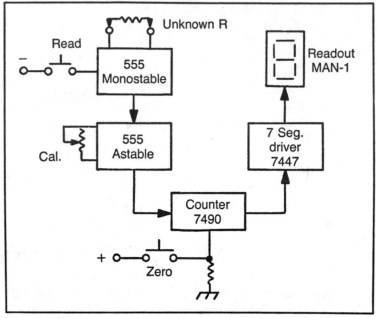

Fig. 10-1. Block diagram of digital ohmmeter.

2. Determine resistance and display it with a single digit, seven-segment readout.
3. Understand the count-reset function and logic of a basic digital ohmmeter.

Materials Required

MAN-1 LED readout (1)
7447 seven-segment decoder driver IC (1)
7490 decade counter IC (1)
555 timer IC (2)
330-Ohm resistor (7)
.1-μF capacitor (1)
1-μF capacitor (1)
1-k ohm resistor (2)
1-megohm resistor (1)
220-k Ohm resistor (1)
470-k ohm resistor (1)
5-megohm potentiometer (1)

Wire the circuit shown in Fig. 10-2. Disconnected wires may

be used for the read and zero switches. Be sure to make *all* connections. Use a red pencil to verify lines while wiring.

When the circuit wiring is complete, momentarily touch pin 2 or 3 of the 7490 to (+). The readout should go to zero (0).

Place a 1-megohm resistor across the unknown R position of the monostable 555. (pins 8 to 6 and 7).

Momentarily touch pin 2 of the monostable 555 to ground (−). The readout should count for a period of time then stop.

Repeat the steps of "zeroing" and "reading" while adjusting the 5-megohm potentiometer until the count advances to 0 (a total count of 10.)

Replace the 1-megohm resistor with the values in Fig. 10-3.

1. Record the count observed with each.

Operation

The counts associated with the resistors in this exercise should indicate the resistance (X 100-k ohm). A full count of 10 digitz - zero to zero - represents 1 megohm. A count of 5 then would

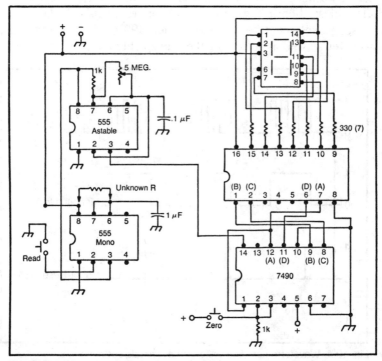

Fig. 10-2. Wiring diagram of 1-digit ohmmeter.

Resistor	Count
220k 470k 1 Meg.	

Fig. 10-3. Table showing relationship between readout and unknown resistance.

represent 0.5 megohm or 500-k ohm. A count of over 10 would be 1 meghom + the reading of the readout. For example, a 2.2-megohm resistor should count to 10 twice and then stop at 2. This represents 1 megohm + 1 megohm + 0.2 megohm or a total of 2.2 megohm. In practice, common 10 percent resistors will yield varied results and this circuit does not provide any great degree of accuracy below 220 -k ohm or above 1 megohm.

In designing a useful ohmmeter, the count-reset function would be automatic rather than manual. This could be done by increasing the speeds of both 555s to eliminate the visible changes of display numbers. In addition, a third 555 in the astable mode could be adjusted for an on-time of short duration (less than 10 percent) and an off time of nearly 100 percent. This would constantly sample the resistor, display its value and reset or recycle. The illustration in Fig. 10-4 shows this relationship. The count-reset pulse would be applied to the 7490 counter to reset the counter as well as to trigger the monostable 555.

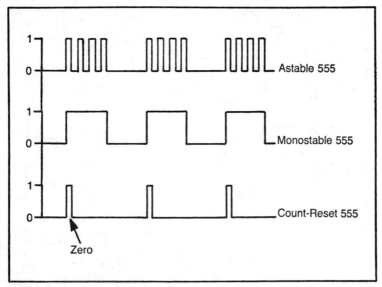

Fig. 10-4. Waveform from 555 units with automatic set-reset 555.

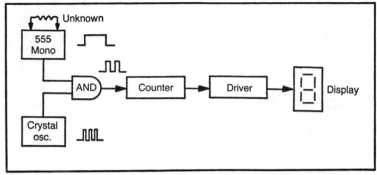

Fig. 10-5. Diagram showing how a crystal time base might be used with a gate to increase the accuracy of the ohmmeter.

This circuit can also be used to determine the value of an unknown capacitor. Since the capacitor used with the 555 also determines the operating frequency, or on-time, in the monostable mode, by fixing the resistance and substituting capacitors, the capacitor value can be read. The capacitor must be completely discharged before measurement is started.

In order to obtain accurate results, it is common to use a crystal time base (clock) in association with the monostable. The illustration in Fig. 10-5 indicates how this arrangement might be developed. Usually a 100-kHz crystal is used since it is a common value and can be obtained easily.

Generally, a one-digit readout is simply not useful for most applications, so digital dividers are driven to allow a number of digits to be displayed at one time. Auto-ranging circuits may also be used to correctly position a decimal point. Additionally, principles such as have been discussed in this exercise are used, along with other principles such as the VCO (voltage-controlled oscillator) to develop sophisticated frequency counters, event timers, and multimeters.

Chapter 11

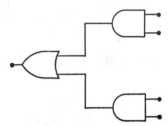

Three Digital Projects

DIGITAL DICE

Many games require some random number-generating system in order to clock or move the players through the steps of the game. Quite often, dice are used for this purpose.

This project provides the complete plans for electronic dice which will provide random numbers in the same fashion as regular hand-thrown dice. The novelty of this circuit is enhanced by its being placed in a case that looks like one large dice.

OPERATION

Figure 11-1 is the schematic diagram of the dice circuit. Two of the NAND gates in IC3 are used as an oscillator which runs when the roll button is pushed. The output of the oscillator feeds IC1, a type 4018 divide-by-six counter. The output of this counter is decoded using a NAND gate and diodes that in turn drive the seven LEDs at the upper left. One output, O5, is used to drive the second type 4018 counter, IC2, which then drives the seven LEDs at the top right.

The LEDs are arranged so that they will light up in the patterns as a die. Since two counters are counting at different speeds, the resulting numbers are random.

The circuit is powered by a 9-volt battery and draws between 15 and 20 mA when operated. If one battery is not sufficient for long use, a second one can be wired in parallel.

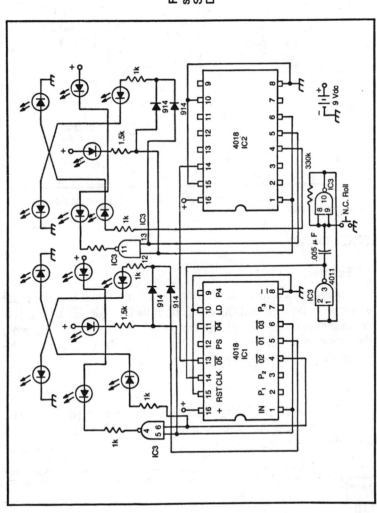

Fig. 11-1. Electronic-dice game schematic. (Courtesy Howard W. Sams Inc., *CMOS Cookbook*, by Don Lancaster.)

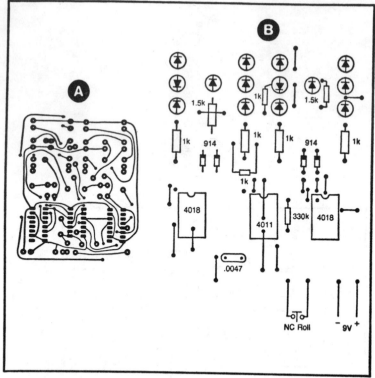

Fig. 11-2. Layout of foil and component sides of the circuit board. (Courtesy Industrial Education.)

CONSTRUCTION

PC Board

The circuit is mounted on a single printed circuit board. Figure 11-2 shows both the foil and component layouts for this board. It is recommended that sockets be used for the ICs in order to avoid damage due to soldering heat. Figure 11-3 shows a view of the completely wired PC board.

Case

The case is made from wood stock, then painted black. The dimensions of the case are given in Fig. 11-4. The lid is sawed off after the case is glued together. This insures that the lid will fit

Adapted from the February 1980 issue of *Industrial Education* magazine with permission from the publisher. Copyright © 1980 by Macmillan Professional Magazines, Inc., 77 Bedford Street, Stamford, CT 06901. All rights reserved.

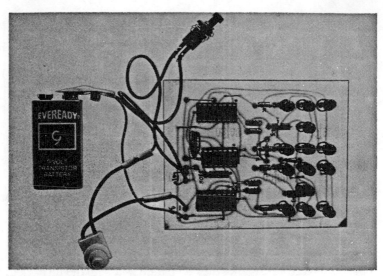

Fig. 11-3. Completed PC board with all wiring done.

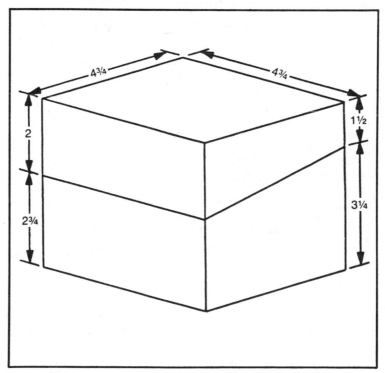

Fig. 11-4. Dice-game case dimensions, not to scale. (Courtesy Industrial Education.)

Fig. 11-5. Completed unit in case. (Photograph by Mary Duncan, Oswego Learning Resources Center. Courtesy Industrial Education.)

properly. Small hinges are mounted at the rear so that the lid may be opened. Figure 11-5 shows the completed unit with the lid closed.

Front Panel

The front panel holds the PC board with the insertion of the LEDs into the proper panel holes. Two switches are located below the dice LEDs. Figure 11-6 is a layout of the front panel. Figure 11-7 shows the completed unit with the lid open. The panel and dice LEDs are clearly shown in this photograph.

Dot Layout

The dots on the case are arranged like those on a common die. Figure 11-8 shows the layout. The dots are ¾ inch in diameter and are made from white contact paper.

Procedure

Follow this procedure when building this project:

1. Secure all parts. Check against the parts list.
2. Lay out and etch the PC board as shown in Fig. 11-2.
3. Drill PC board.
4. Mount jumper wires and sockets; then solder each.
5. Mount LEDs so that each is well above the rest of the components. Solder each using a heat sink.
6. Install and solder all resistors and other components.
7. Mount ICs in sockets, observing correct pin layout.
8. Attach switches and battery; then test circuit.
9. If the circuit operates, go on to the next step. If it does not

function correctly, troubleshoot until it operates. Check battery, soldering, etc.

10. Construct the case according to Fig. 11-4.
11. Saw off lid, install hinges, sand, then finish the case with black paint.
12. Make white dots and attach to case according to the pattern in Fig. 11-8. Make sure the dots are cemented flat.
13. Overspray the dots with a clear finish so they will not rub off during use.
14. Construct the front panel according to Fig. 11-6.
15. Paint the front panel white.
16. Black out the area around the LED holes.
17. Rub on legends at the switch openings.
18. Install switches.
19. Install the PC board to the front panel by pressing the LEDs through the die holes. The LEDs should protrude

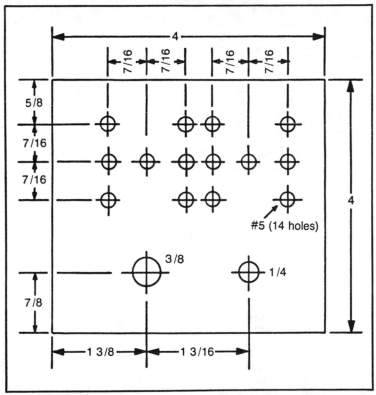

Fig. 11-6. Layout of front panel. (Courtesy Industrial Education.) (Not to scale.)

Fig. 11-7. Completed unit in place in the case. (Photograph by Mary Duncan, Oswego Learning Resources Center. Courtesy Industrial Education.)

about one-eighth inch above the panel surface. All LEDs should protrude the same amount.

20. Glue side rails of wood around the inside of the case.
21. Press-fit the panel, with the battery attached, into the case. If the fit is loose, use screws to secure.
22. Operate the unit.

PARTS LIST

Item	Description	Quantity
1	LED, red	14
2	Resistor, 1000 ohms	6
3	Resistor, 1500 ohms	2
4	Diode, general purpose type 914	4
5	Resistor, 330-k ohms	1
6	Ceramic disk capacitor, .0047 μF	1
7	Switch, NO push type (roll)	1
8	Switch, SPST push type (on/off)	1
9	Battery, 9 V	1
10	Battery snap	1
11	Socket, 14-pin DIP	1
12	Socket, 16-pin DIP	2

13	IC, type 4011 quad NAND	1
14	IC, type 4018 counter	2
15	Miscellaneous screws, PC stock, wood, glue, paint, wire, solder, hinges, etc.	•

"CLIMB THE MOUNTAIN"

"Climb-the-Mountain" is a two-player digital game which is based only on luck. No real skill which would give one player an advantage can be developed. This project will provide the basic theory of how the game works, followed by the complete construction details.

OPERATION

Each player uses a circuit such as the one shown in Fig. 11-9. The circuit works in the following way. The timer, IC1, when started by the "roll" button, produces pulses at about a 1000-Hz rate. These pulses are counted by IC2, the decade counter. The decade counter provides a BCD output to the input of IC3.

IC3 is a one-of-ten decoder. As the BCD count changes, each of its ten output pins goes to ground, zero logic, in sequence. The other nine are positive, or high.

When the "roll" button is released, one of the ten outputs from IC3 will be low, or ground. The outputs are connected through

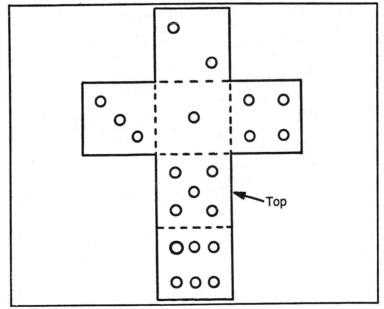

Fig. 11-8. Layout of die markings used on case.

145

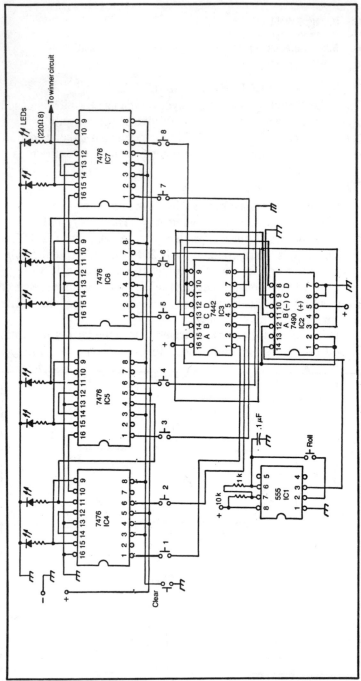

Fig. 11-9. Schematic of one player circuit for Climb-the-Mountain. Two identical circuits are needed.

push-button switches to a series of flip-flops in a shift register, ICs 4 to 7.

The eight flip-flops of the shift register are connected so that each can be activated only if two conditions are present: the preceding flip-flop must be on and its switch must be grounded by IC3.

Since only eight switches are provided, the 8, 9, and 10 outputs of IC3 are tied together to the number 8 switch. This provides better odds for this switch. It is the last in the series before reaching the top and winning the game.

One player "rolls" an IC1 and presses the number-1 button. If the number-1 switch happens to be grounded by IC3 at the time it is pressed, the first LED will light. The player has three "rolls," then it is the other player's turn. When a player lights the first lamp, the number-2 button is pressed after the next "roll." The lights must be lighted in sequence up the mountain until LED number 8 is lighted.

Winner Circuit

When the final LED is lighted by one player an alarm beeper will sound and a flashing LED will indicate the winner. Figure 11-10 is the schematic diagram of this circuit. It consists of two complete "beep" alarm circuits using type 555 timer ICs.

CONSTRUCTION

Figure 11-11 is the foil pattern for the player PC boards. Two of these boards are required. Surplus calculator buttons are used for push switches on this board. They are numbered from 1 to 8 with rub-on lettering. The original lettering can be scraped away with abrasive paper. Figure 11-12 shows the component layout of the PC boards.

Figure 11-13 is the foil layout for the alarm PC board. Figure 11-14 is the component view of this board. All ICs are mounted in sockets for easy troubleshooting and to avoid heat damage during soldering.

Figure 11-15 shows the interconnections between the alarm board and the two player PC boards. Notice that a 6-volt battery is used to power the circuit. A regulated power supply could be constructed to replace the battery if desired.

Case

The case is constructed of wood and hardboard. The dimensions of the base are given in Fig. 11-16. The player boards are

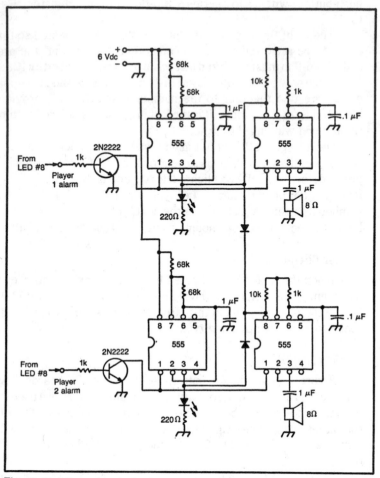

Fig. 11-10. Winner alarm circuit.

mounted so that the keyboards are visible in front of the "mountain."

Figure 11-17 shows the approximate dimensions of the mountain display which contains the LEDs. This display should be painted black so that the LEDs will show better when lighted.

Assembly

Figure 11-18 shows the completed unit with the mountain in place, ready for play. Figure 11-19 shows the rear view of the unit. Notice that the LEDs are wired to each player board. Figure 11-20 shows the under side of the completed unit. The battery is located at the bottom left. The alarm board and speakers are located at the right.

Figure 11-21 is a close-up view of the alarm board. The speakers are held to the wood with wood screws and washers.

PROCEDURE

Follow this step-by-step procedure during construction of this project:

1. Secure all parts. Check against the parts list.
2. Fabricate two player PC boards and one alarm PC board. Use the usual method for layout, etching, and drilling.
3. Mount components on each board and solder all joints except the interconnections.
4. Construct the base according to the layout diagrams.
5. Paint and apply lettering to the deck.

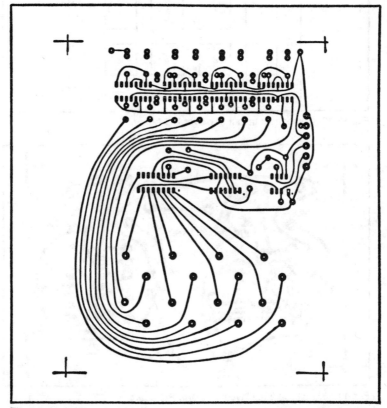

Fig. 11-11. Half-scale foil layout for PC board.

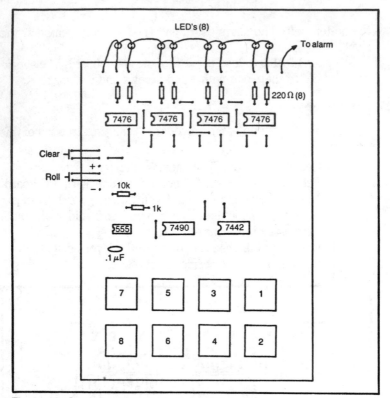

Fig. 11-12. Component layout of player PC board.

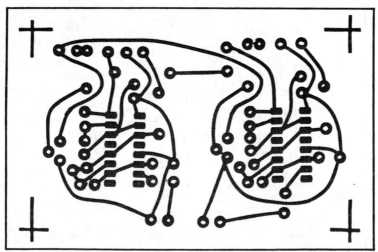

Fig. 11-13. Full-scale foil layout of alarm PC board.

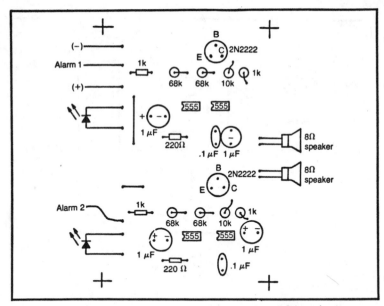

Fig. 11-14. Full-size component layout of alarm winner PC board.

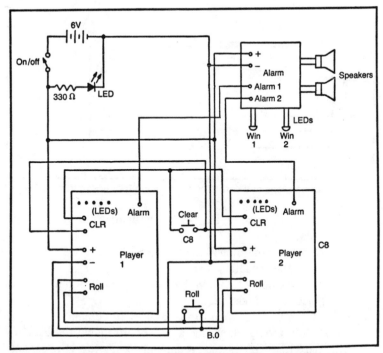

Fig. 11-15. Interconnection diagram for Climb-the-Mountain game.

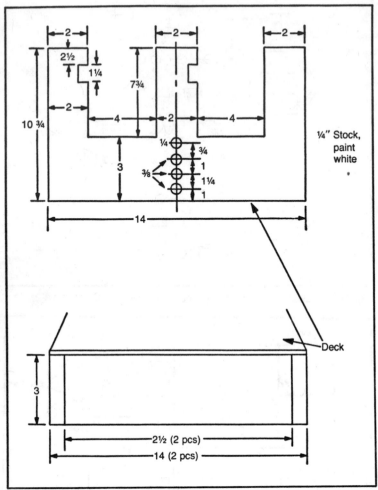

Fig. 11-16. Dimensions and details of construction for deck (not to scale).

6. Install the player PC boards with machine screws and nuts.
7. Construct the mountain display.
8. Paint the mountain black.
9. Mount display LEDs in the mountain and cement them in place with model glue.
10. Attach the mountain to the base with cement.
11. Mount the speakers and alarm board inside the base.
12. Wire LEDs on the mountain, and complete all the connections according to the wiring diagram.

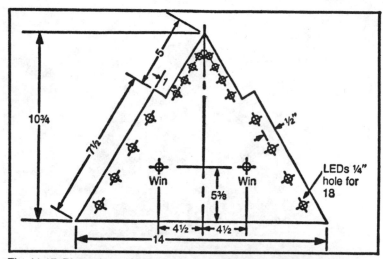

Fig. 11-17. Dimensions of "mountain" background for game (not to scale).

Fig. 11-18. Completed game ready to be operated.

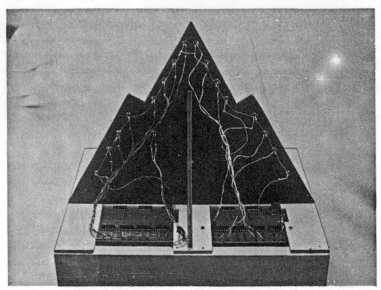

Fig. 11-19. Rear view of unit showing the back of the "mountain" with the wiring to the LEDs. The player boards are visible below.

13. Attach the battery and test the unit.
14. Construct sides and rear of the mountain from heavy cardboard painted black. Cement in place.

Fig. 11-20. View of underside of completed unit. The alarm circuit is located at the right on the case wall.

Fig. 11-21. Close-up view of the alarm/winner PC board and speakers mounted to the inside of the case. Speakers are secured with screws and washers. Speakers may be located at any other convenient place.

GAME RULES

The following rules are suggested for playing Climb-the-Mountain. They may be changed as the players desire.

Object of Game

1. The object of the game is to light the eight lamps that lead to the top of the mountain. The lamps must be lighted in order, starting from the bottom of the mountain. The bottom lamp is number one.
2. The players take turns trying to light their lamps. The first player to light all eight lamps is the winner.
3. The losing player has first turn in the next game.

Starting a Game

1. Turn the power on. The "ready" lamp will light.
2. Press the reset button to turn off all lamps on the mountain.

The Play

1. The first player presses the "roll" button and releases it after a few seconds.
2. After releasing the roll button, the player presses button number one. If the player is lucky, the first lamp will light. There is a 10% chance of success.

3. The player may roll three times. After three rolls, it becomes the other player's turn.
4. A player that lights a lamp attempts to light the next lamp up the mountain at the next turn. To light the second lamp, press the roll button as before; then press button number two, etc.
5. Players continue taking turns until one is the winner.
6. Press the reset button to start a new game.

PARTS LIST

Item	Description	Quantity
1	IC timer, type 555	6
2	Speaker, 8 ohms, 2-inch diameter	2
3	Transistor, NPN, type 2N2222	2
4	Resistor, 68-k ohms	4
5	Resistor, 10 ohms	4
6	Resistor, 1000 ohms	6
7	Capacitor, 1 μF	4
8	Resistor, 220 ohms	19
9	Capacitor, .1 μF	4
10	LED, red	19
11	IC, type 7476 JK flip-flop	8
12	IC, type 7490 decade counter	2
13	IC, type 7442 decoder	2
14	Push switch, calculator type, NO	16
15	Push switch, NO (clear and roll)	2
16	Lantern battery, 6 V	1
17	Socket, 8-pin DIP	2
18	Socket, 14-pin DIP	2
19	Socket, 16-pin DIP	12
20	Switch, SPST, push type, on-off	1
21	Miscellaneous wire, solder, PC stock, wood, bristol board, nuts, screws, paint, glue, etc.	

14-NOTE MUSIC GENERATOR

This project is a tone generator which can be programmed to play a sequence of 14 notes. The sequence can be random tones or the unit can be adjusted to play a tune. After the unit is started manually, it will complete the sequence and reset itself.

THE CIRCUIT

Figure 11-22 is the complete schematic diagram of this circuit. Although it looks complicated, it is really quite simple. Consider each part of the circuit in turn, and see how each is related to the

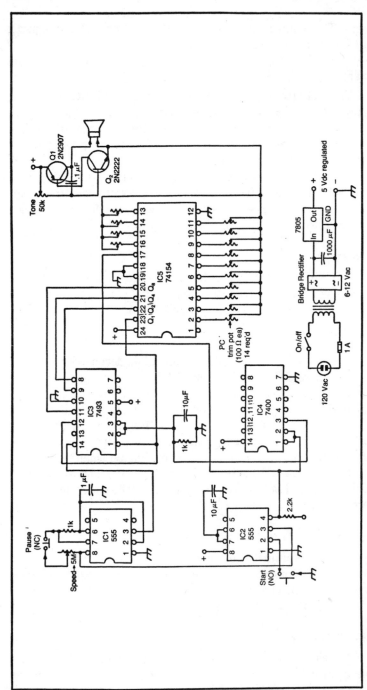

Fig. 11-22. Schematic of 14-note music generator.

157

others. The following information will show how each of the parts works within the complete circuit.

Timers

There are two type 555 timer ICs in this circuit. IC1 is used to supply a series of pulses for the unit. Each time one of these pulses is produced, the tone generator goes to the next note. The push-button switch labelled "pause" allows the operator to stop the pulses so that one note continues.

IC2 is wired as a switch. It is used to activate the unit and to turn it off at the end of the sequence. IC2 goes into operation when the start button is pressed. It stops the sequence when it receives a stop pulse from IC5, pin 17.

Counter

IC3 is a 16-bit binary counter. It will provide a BCD output at terminals 8, 9, 11, 12 in response to pulses at input pin 14. When IC is set to zero, terminals 8, 9, 11, and 12 are all low.

BCD Decoder

The BCD code which comes from IC3 is decoded by IC5. For each step of the BCD input, the next of the sixteen output terminals goes to ground. (The output terminals are 1 through 11 and 13 through 17.) The other fifteen output terminals are high.

When the circuit is at rest, terminal 1 of IC5 is low. When the start button is pressed, the first pulse from IC1 causes terminal 2 of IC5 to go low and produce the first note. Finally, when terminal 17 goes low, it stops IC2 and resets IC3 back to the rest position. This causes terminal 1 of IC5 to return to low, and the entire unit is once more at rest.

Reset IC

The output from IC5 terminal 17 is a low, or ground. This is the correct polarity of stop signal for IC2, but a positive reset voltage is required for IC3. The function of IC4 is to invert the low stop signal from IC5 into a positive reset pulse for IC3.

Audio Oscillator

The audio oscillator consists of two transistors, Q1 and Q2. A tone control is used to set the overall range of the tones it can produce. The potentiometer which is grounded by IC5 at any moment also controls the frequency of the oscillator. Each of these potentiometers can be adjusted to produce a specific tone, so that tunes can be played. The potentiometer connected to terminal 2 of IC5 controls the tone of the first note. The one connected to terminal 16 controls the last one.

Power Supply

A 5-volt regulated power supply is provided for this unit. It is mounted on a separate PC board. It uses a 12-volt transformer, bridge rectifier, and solid-state regulator.

Start and Stop

To start the unit, IC2 is activated by grounding pin 2 with the push-button switch. This, in turn, activates IC1, and it will begin to send pulses to IC3.

The pulses received by IC3 produce a BCD output, which is applied to the input of IC5. The BCD sequence causes the output pins of IC5 to go to ground, each in turn. As each output terminal goes to ground, a new note comes from the oscillator.

The last output, pin 17, is not hooked to the oscillator, but is used to reset the unit back to its starting position. This pulse is applied to IC2, pin 4, which resets it, or turns it off. This stops IC1 and no more pulses are sent forward until the unit is once again activated.

This same pulse that stops IC2 is inverted by IC4 and applied to IC3. This causes it to clear back to its starting point, or zero.

The entire unit is self-completing in its sequence of notes. It does not matter if the sequence is slow or fast. It will complete itself and reset to stop; then it will wait for the next start command from the user.

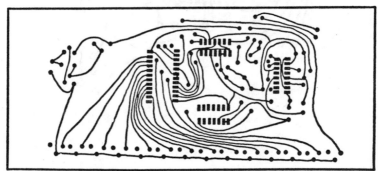

Fig. 11-23. Foil layout of PC board.

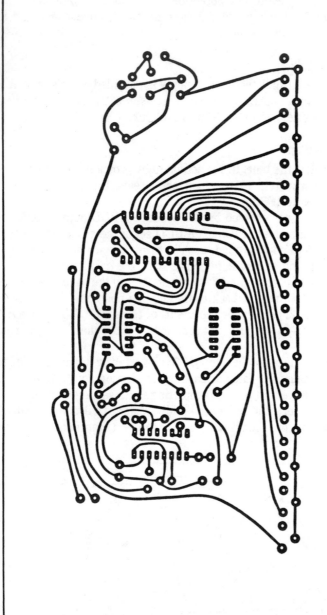

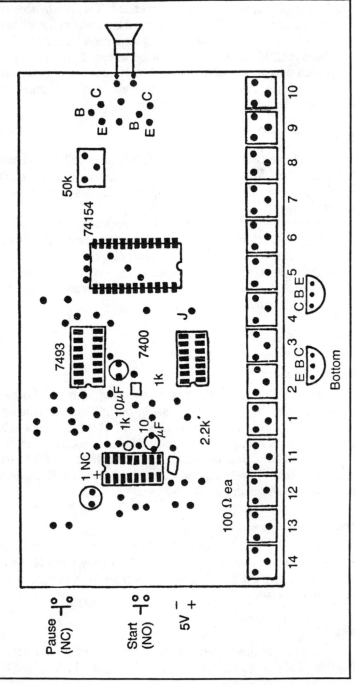

Fig. 11-24. Component layout of main PC board.

161

IC5 remains at rest with pin 1 grounded. This is why the first note position is not used. Since the first and last output pins are used for rest and reset, only 14 of the 16 can be used for notes.

The speed of the sequence can be adjusted by varying the 5-megaohm potentiometer connected to IC1. A normally closed, push-button pause switch is used for manual operation. This allows each note to be adjusted as desired.

USE

This device can be used as an unusual door bell or just as an attention-getter. When random notes are played in a fast mode, unusual "computerlike" sounds are generated. This may be useful for sound effects.

The sound level is about right for a normal room. It may be reduced by connecting a resistor in series with the speaker. If more volume is needed, the unit can be attached to an amplifier.

Modifications can be made to this unit to extend the sequence. A flip-flop could be used to step to a second decoder and add another 14 notes, for example.

CONSTRUCTION

Main PC Board

Figure 11-23 is the foil pattern for the main PC board. Sockets are used for the ICs. Note that IC5 is a 24-pin DIP unit, and the size of the socket for it is larger than the others. This board should be made in the usual manner. See Chapter 3 for details on PC boards.

Figure 11-24 shows the component side of the main PC board. Be sure to install the jumpers under the socket of IC5 before soldering the socket in place.

Power Supply

A small PC board is used for the rectifier, capacitor, and regulator of the power supply. Figure 11-25 shows the layout of both sides of this PC board. The transformer is mounted on the bottom of the case.

Case

The case for this project is made from ⅛-inch clear acrylic plastic so that the PC boards can be seen. Figure 11-26 gives the

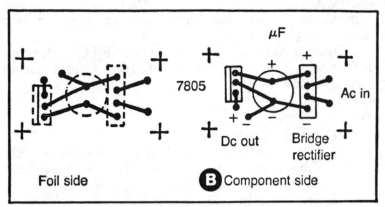

Fig. 11-25. Full-scale layouts of power-supply PC board (not to scale).

approximate dimensions for this case. It can be made smaller if it is desired. Stick-on feet are attached to the bottom of the case at the corners.

Figure 11-27 shows the completed unit. The speaker is cemented to the inside of the top. The start and pause buttons are located on the top. A remote start jack is located on the back. This allows a remote start switch to be used with the unit. The PC boards

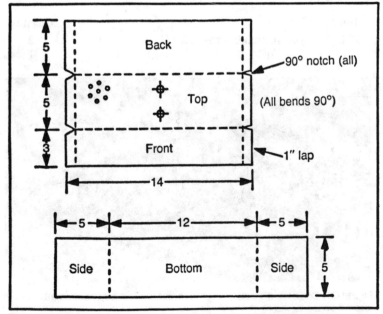

Fig. 11-26. Approximate layout of acrylic case (not to scale).

and transformer are bolted to the base. The top is held to the base with machine screws. Holes in the base are tapped for these screws.

Figure 11-28 shows a closer view of the main PC board. Access to the row of potentiometers is provided. These are used to adjust each of the notes.

Procedure

1. Secure all parts.
2. Fabricate the PC boards for power supply and main circuit.
3. Fabricate the case. Be sure to drill all holes before bending the plastic to final shape.
4. Mount switches and transformer.
5. Wire power supply, main PC board, and other components according to the schematic diagram.
6. Cement speaker to case top.
7. Finish assembly of case.
8. Fasten stick-on feet to base corners.
9. Attach power cord.
10. Test unit and use adjustments to provide different speeds and tones.

PARTS LIST

Item	Description	Quantity
1	IC, timer, type 555	2
2	Potentiometer, PC type, 5 megohms	1

Fig. 11-27. Completed 14-note music generator. Start and pause switches are located on top to the right of the speaker.

Fig. 11-28. Close-up of completed music generator. Potentiometers facing out are used to adjust each of the notes.

3	Switch, push-button NO	1
4	Switch, push-button NC	1
5	Socket, 8-pin DIP	2
6	Socket, 14-pin DIP	2
7	Socket, 24-pin DIP	1
8	IC, quad NAND gate, type 7400	1
9	IC, counter, type 7493	1
10	Capacitor, 10 μF, 10 V	2
11	Capacitor, 1 μF, 10 V	1
12	Speaker, 8 ohms, 2-inch	1
13	Resistor, 1000 ohms, ¼ W	2
14	Resistor, 2200 ohms, ¼ W	1
15	Potentiometer, PC type, 50-k ohms	1
16	Potentiometer, PC type, 100-k ohms	14
17	Transistor, NPN, type 2N2222	1
18	Transistor, PNP, type 2N2907	1
19	IC decoder, type 74154	1
20	Transformer, 120 Vac to 12 Vac	1
21	Line cord with plug	1
22	Capacitor, 1000 μF, 20 V	1
23	Bridge rectifier, 20 V, 1A	1
24	Regulator, 5 V, type 7805	1
25	Switch, on-off SPST (Optional)	1
26	Fuse and holder, 1A	1
27	Miscellaneous nuts, bolts, wire, solder, acrylic plastic, etc.	-

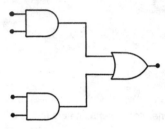

Answer Key

CHAPTER 5: LOGIC GATES

Exercises

1.
C
1
1
1
0

Fig. 5-3.

2. $C = \overline{A \cdot B}$ or $\overline{C} = A \cdot B$

3. This is the NAND function.

4.
C	D
1	0
1	0
1	0
0	1

Fig. 5-5.

5. $C = A\overline{B}$ or $\overline{C} = A B$
   $D = A B$

6. This is an AND function.

   Fig. 5-8   Fig. 5-10

7.
B
1
0

B
1
0

8.
C
1
0
0
0

Fig. 5-12.

9. $C = \overline{A} + \overline{B}$ or $\overline{C} = A + B$

10. This is a NOR function.

11.
C	D
1	0
0	1
0	1
0	1

Fig. 5-14.

12. $D = A + B$

13. OR

14.
C
0
1
1
1

Fig. 5-17.

15. $A + B$

16. OR
17.

C
0
0
0
1

18. AB
19. AND
20.

D	E
1	0
1	1
1	0
1	0
1	0
1	1
0	1
0	1

21. $D = \overline{A}\,B$
   $E = (\overline{A} + \overline{B}) + C$
22. $D = (A + B)\,C$
23. A. $A + BC = D$

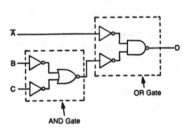

AND Gate   OR Gate

B. $A\overline{B} + \overline{C}D = E$

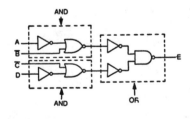

AND   AND   OR

C. $AB + C = \overline{D}$

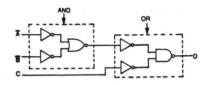

AND   OR

Quiz

1. C
2. A
3. B
4. B
5. A
6. E
7. E
8.

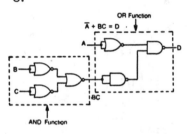

$\overline{A} + BC = D$ OR Function

AND Function   BC

9.

C
1
1
1
0

10.

C
1
0

0	0

CHAPTER 6: FLIP FLOP

Exercises

1. $Q = 1$
   $\overline{Q} = 0$
2. $Q = 0$
   $\overline{Q} = 1$
3. $Q = 0$
   $\overline{Q} = 1$
4. $Q = 1$
   $\overline{Q} = 0$
5. $Q = 0$
   $\overline{Q} = 1$
6. $Q = 1$
   $\overline{Q} = 0$
7. Q $\overline{Q}$
   5   5
8. No

9.

A	B	C
1	1	1
0	1	1
1	0	1
0	0	1
1	1	0
0	1	0
1	0	0
0	0	0

Fig. 6-9.

10. Yes
11. Yes

12. $A = 4 \times C$ $(C = \dfrac{A}{4})$

   (A divide-by-4 Counter)

13. $\boxed{\begin{matrix} B \\ 5 \end{matrix}}$ Fig. 6-11.

14. $A = 2 \times B$
15. Division (by 2)

Quiz

1. D

2. B
3. E
4. E
5. A
6. D
7. B
8. E
9. B
10. A

CHAPTER 7: DIGITAL COUNTERS AND DECIMAL DISPLAYS

Exercises

1.

D	C	B	A
0	0	0	0
0	0	0	1
0	0	1	0
0	0	1	1
0	1	0	0
0	1	0	1
0	1	1	0
0	1	1	1
1	0	0	0
1	0	0	1

Fig. 7-2.

2. The counts goes to 0000 and starts over again, counting upwards.
3. Decimal
   0
   1
   2
   3
   4
   5   Fig. 7-5.
   6
   7
   8
   9
4. No
5. It counts downwards to zero.

Quiz

1. C
2. B
3. B
4. E
5. D
6. C
7. E
8. C
9. E
10. B

6. No

Quiz

1. B
2. A
3. B
4. C
5. D

CHAPTER 8: SHIFT REGISTERS

Exercises

1.

A	B	C	D
1	0	0	0
0	1	0	0
0	0	1	0
0	0	0	1
0	0	0	0

Fig. 8-2.

2. The pulse shifted from flip-flop to flip-flop to the right until the data was shifted out and the register remains at 0000.

3. 4

4.

D	C	B	A
0	0	0	1
0	0	1	0
0	1	0	0
1	0	0	0
0	0	0	1
0	0	1	0
0	1	0	0
1	0	0	0
0	0	0	1
0	0	1	0
0	1	0	0

Fig. 8-3.

5. The shift continued over and over again due to one shift from D back to A.

CHAPTER 9: THE EXCLUSIVE - OR GATE

Exercises

1. C
   0
   1   Fig. 9-2.
   1
   0

2. The OR gate table provides a 1 output when *either* input is a 1. The exclusive-OR gate provides a 1 output *only* when the two inputs are at different logic states.

3. F
   1
   0
   0
   0
   1   Fig. 9-4.
   1
   0
   1
   1

4. When an odd number of 1s appear in ABCDE, the F output is always a 1. When an even number of 1s appear in ABCDE, then F is always a 0.

5. 1

6. 0
7. C
   0
   0
   0
   0   Fig. 9-6.
   1
   1
   1
8. 0
9. 1

Quiz

  1. D

2. B
3. C
4. A
5. E
6. E

CHAPTER 10: A DIGITAL SYSTEM

Exercise

  1. Count
     2 - 3
     4 - 5   Fig. 10-3.
     9 - 0

Appendix:

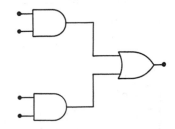

5400/7400 TTL ICs

The following short-form specifications of a selected number of 54/74 series TTL ICs is provided so that you may have a quick reference to each. In each case, a pin diagram is provided with a brief explanation of the use and/or special characteristics of the unit.

The information contained in this appendix has been adapted from publications produced by Texas Instruments Inc. This material is used through the courtesy of, and with the permission of, Texas Instruments Inc., Dallas, Texas. For a more extensive coverage of the specifications of these and other ICs, you should consult a good TTL reference book such as *The TTL Data Book for Design Engineers,* Second Edition, Texas Instruments Inc., 1976.

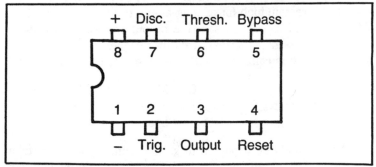

Fig. A-1. Pin diagram of 555 Timer.

555 TIMER

The 555 timer IC is very popular as an inexpensive and accurate clock (astable) or one-shot monostable unit. It is supplied in an 8-pin DIP package and can source or sink approximately 200 mA. Both monostable and astable wiring and information are provided below.

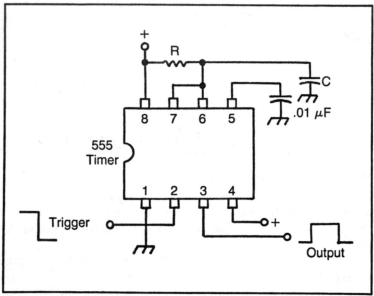

Fig. A-2. Monostable connections.

Operation

When a negative pulse is applied to the trigger (pin 2), the output (pin 3) goes from its resting level (low) to high (+). The output will remain high for as long as the timing network allows. After this time period passes, the output will return to its low condition and remain there until another trigger pulse is provided. The "on" time is calculated with the following equation:

$$T = 1.1\ RC$$

C = Microfarads (μF).
R = Megohms.
T = Seconds.

Pin 5 should be bypassed to ground with a .01 μF tantalum capacitor. The circuit may work if this is not done, but most literature recommends bypassing.

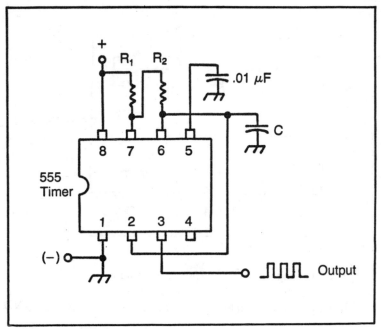

Fig. A-3. Astable wiring connections.

When wired in the astable or free-running mode, the frequency of the output is calculated with the following equation.

$$f = \frac{1.5}{(R_1 + 2R_2)C}$$

f = Hertz.
R = Megohm.
C = Microfarad (μF).

If R_2 is large when compared to R_1, the output square wave will be on and off about an equal amount of time. The on and off time relationship can be altered by changing the ratio between these two resistors.

Pin 5 should be bypassed to ground with a .01 μF tantalum capacitor. The circuit may work without this bypass capacitor but it is recommended as a good practice.

7400 QUAD 2-INPUT POSITIVE NAND GATE

Features

Each gate can be used separately.
5-volt supply must be used.

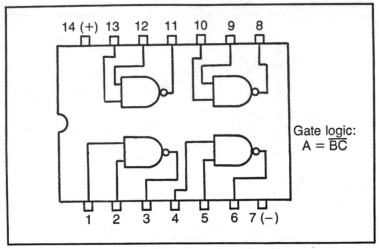

Gate logic:
$A = \overline{BC}$

Fig. A-4. 7400 pin diagram.

Total package current is 20 mA maximum.
This package is easy to use and costs very little.
It is frequently used as a source of inverters as well as NAND gates.

7402 QUAD 2-INPUT POSITIVE NOR GATES

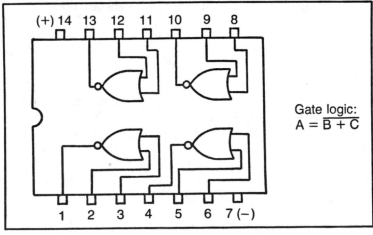

Gate logic:
$A = \overline{B + C}$

Fig. A-5. 7402 pin diagram.

Features

5-volt supply must be used.

Each gate can be used separately.
Package current is 20 mA maximum.

7404 HEX INVERTER

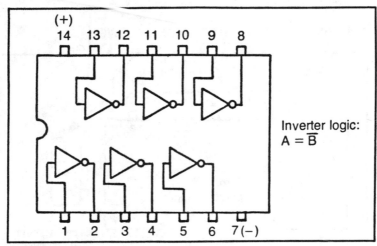

Fig. A-6. 7404 pin diagram.

Features

Use 5-volt supply only.
Each inverter can be used separately.
Package current is 20 mA maximum.

7432 QUAD 2-INPUT OR GATES

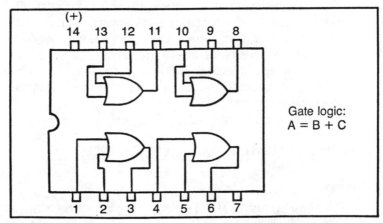

Fig. A-7. 7432 pin diagram.

Features

Use 5-volt supply only.
Package current is 20 mA.
Each gate can be used separately.

7447 BCD-TO-SEVEN-SEGMENT DECODER/DRIVER

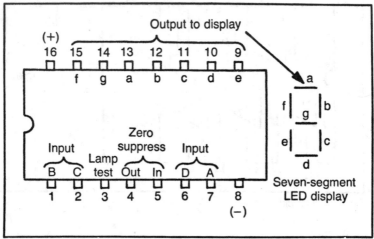

Fig. A-8. 7447 pin diagram.

Features

Use 5-volt supply only.

Output lines must have a series resistor (330 ohms) when driving a common-anode seven-segment LED display. Outputs are low (−) for operating and can sink about 40 mA.

Pin 3, lamp test, normally remains high (+). If it is made low (−), all segments of the display will light.

Zero blanking will occur if pin 5 is made low (−). Pin 4 is used to pass on a ground (−) to previous units for leading zero suppression.

Package current is approximately 64 mA.

7476 DUAL J-K FLIP-FLOPS WITH CLEAR AND PRESET

Features

Use 5 volts only. NOTE: Power supply pins are not located at the usual place but are pins 5 and 13.
Each flip-flop can be used separately.

176

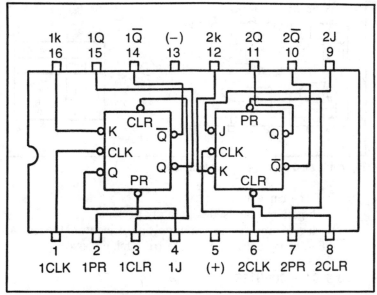

Fig. A-9. 7476 pin diagram.

Package current is 40 mA.

Maximum operating frequency is 20 MHz.

When Clear (CLR) is made low (−), output will go to Q=0, $\overline{Q}$=1.

When Preset (PR) is made low (−), outputs will go to Q = 1, $\overline{Q}$ = 0.

Do not make both Clear and Preset low at the same time or output control will be lost.

If J=0 and K=0, no change in output will occur when a clock pulse is received. Flip-flops operate only on a clock pulse. J and K must be changed after the pulse.

If J=1 and K=0, the output will go to Q=1, $\overline{Q}$=0 when a clock pulse is received.

If J=0 and K=1, the clock pulse will cause the output to go to Q=0, $\overline{Q}$=1.

If both J and K are high (+), the output will toggle or divide the input frequency by 2.

7486 QUAD 2-INPUT EXCLUSIVE-OR GATES

Features

Use 5 volts only.

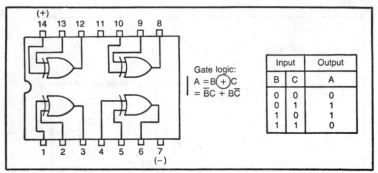

Fig. A-10. 7486 pin diagram and truth table.

Package current is 30 mA.

Each gate can be used separately.

Each gate provides an output only if the inputs of the gate are different logic levels. See the truth table above.

7490 DECADE COUNTER

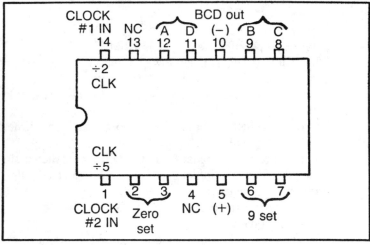

Fig. A-11.7490 pin diagram.

Features

Use only 5 volts. NOTE: Power connections are to pins 5 and 10.

Maximum frequency is 16 MHz.

Two separate counters are in this package. The divide-by-two input is at pin 14 and the divide-by-five input is pin 1.

In order to arrange a divide-by-10 counter, the output of D (pin 11) can be jumpered to clock 1 (pin 14) and pin 1 becomes the clock input. A BCD output results from this configuration.

The counter can be set to zero by making pin 2 or 3 (or both) positive (+). The counter can be set to 9 by bringing both (or either) pins 6 and 7 to a positive (+) state.

The input clock signal must be bounceless.

All set pins (0 and 9) should be held at ground (−) during operation.

The counter ripple counts only in the up direction.

74192 SYNCHRONOUS BCD UP/DOWN CLOCK COUNTER WITH CLEAR AND PRELOAD

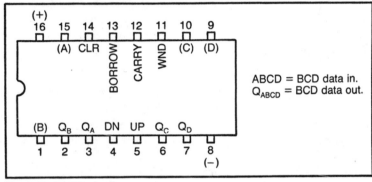

Fig. A-12. 74192 pin diagram.

Clock input is attached to either up or down, depending on desired direction. The other pin not used for the clock must be held at positive (+) during counting. The output Q_{ABCD} is a BCD code.

Data can be "loaded" into the counter by applying the correct BCD code to the input pins (ABCD) and momentarily bringing the load (pin 11) low (−). The load pin must remain high (+) during the time the unit is counting.

The unit can be cleared to zero by momentarily bringing the clear (pin 14) to a high (+) state. The clear pin must be returned to low (−) in order to count.

Units can be cascaded with the carry and borrow outputs used to clock to or from other units.

Use only 5 volts.

Package current is 65 mA.

Maximum frequency is 32 MHz.

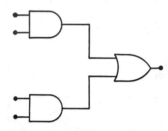

Bibliography

Boyce, Jefferson, *Digital Logic and Switching Circuits*. Englewood Cliffs, NJ: Prentice-Hall, Inc., 1975.

Dempsey, John A., *Experimentation with Digital Electronics* (Laboratory Manual). Reading, MA: Addison-Wesley Publishing Co., 1977.

Doyle, John M., *Digital Switching and Timing Circuits*. North Scituate, MA: Duxbury Press, 1976.

Floyd, Thomas L., *Digital Logic Fundamentals*. Columbus, OH: Charles E. Merrill Publishing Co., 1977.

Gothmann, William H., *Digital Electronics. An Introduction to Theory and Practice*. Englewood Cliffs, NJ: Prentice-Hall, Inc., 1977.

Hawkins, H., *Digital Electronics Projects*. Blue Ridge Summit, PA: TAB BOOKS Inc., 1983.

Kershaw, John D., *Digital Electronics: Logic and Systems*. North Scituate, MA: Duxbury Press, 1976.

Lancaster, Don, *TTL Cookbook*. Indianapolis, IN: Howard Sams & Co., Inc., 1974.

   CMOS Cookbook. Indianapolis, IN: Howard Sams & Co., Inc., 1977.

Leach, Donald P., *Experiments in Digital Principles* (Laboratory Manual). New York, NY: McGraw-Hill Book Co./Gregg Division, 1976.

Levine, Morris E., *Digital Theory and Practice Using Integrated*

Circuits. Englewood Cliffs, NJ: Prentice-Hall, Inc., 1978.

    Digital Theory and Experimentation Using Integrated Circuits (Laboratory Manual). Englewood Cliffs, NJ: Prentice/Hall, Inc., 1974.

Malvino, Albert and Leach, Donald, *Digital Principles and Applications* (Second Edition). New York, NY: McGraw-Hill Book Co., 1975.

New York Institute of Technology, *A Programmed Course in Basic Pulse Circuits*. McGraw-Hill Book Co./Gregg Division, New York, NY: 1978.

Porat, Dan and Barna, Arpad, *Introduction to Digital Techniques*. New York, NY: John Wiley and Sons, 1979.

Rhyne, Thomas V., *Fundamentals of Digital Systems Design*. Englewood Cliffs, NJ: Prentice-Hall, Inc., 1973.

Rutkowski, George and Olesky, Jerome, *Fundamentals of Digtial Electronics* (Laboratory Text). Englewood Cliffs, NJ: Prentice-Hall, Inc., 1978.

Sandige, Richard S., *Digital Concepts Using Standard Integrated Circuits*. New York, NY: McGraw-Hill Book Co., 1978.

Williams, Gerald E., *Digital Technology, Principles and Practices*. Chicago, IL: Science Research Associates, Inc., 1977.

    Digital Technology (Laboratory Manual). Chicago, IL: Science Research Associates, Inc., 1977.

Wojslaw, Charles, *Integrated Circuits: Theory and Applications*. Reston, VA: 1978. Publishing Co., Inc., 1978.

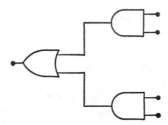

Suppliers

The following companies sell electronic components by mail.

Active Electronics Sales Corp.
P.O. Box 1035
Framingham, MA 01701
Excellent quantity-prices. Wide variety of ICs and other components.

Advanced Computer Products Inc.
P.O. Box 17329
Irvine, CA 92713
Solder, IC breadboards, wire-wrap tools, PC-board layout materials, cases, components, and parts such as resistors, etc.

Allied Electronics
401 East Eighth Street

Fort Worth, TX 76102
Wide variety of electronic parts, hardware, components, and vacuum tubes. PC-board supplies. Quantity prices.

Arch Electronics Co.
1318 Arch Street
Philadelphia, PA 19107
Variety of components. Vacuum tubes. Quantity prices.

B & F Enterprises
119 Foster Street
Peabody, MA 01960
Wide variety of components and hardware. Many unusual items.

Burstein - Applebee
3199 Mercier Street

Kansas City, MO 64111
Wide variety of electronics tools, equipment, and components.

Contact East Inc.
7 Cypress Drive
Burlington, MA 01803
Excellent selection of tools and equipment for PC board and electronics use. Production equipment and supplies.

Digi-Key Corp.
P.O. Box 677
Highway 32 South
Thief River Falls, MN 56701
Excellent source of ICs, capacitors, resistors, PC-layout materials. Good quantity prices.

Electronic Supermarket
P.O. Box 619
Lynnfield, MA 01940
Hardware, test equipment, PC boards, transformers, and many other items.

Fordam Radio
855 Conklin Street
Farmingdale, NY 11735
Test equipment, solder supplies, repair parts for Radio and TV.

Hanifin Electronics Corp.
P.O. Box 188
Bridgeport, PA 19405
Wide variety of semiconductors and other components.

Herbach & Randeman Inc.
401 East Erie Avenue
Philadelphia, PA 19134
Motors, fans, test equipment, and many unusual items. Catalog features special items each month.

J. Meshna
P.O. Box 62
E. Lynn, MA 01904
Surplus and unusual items. Wide variety with many bargain prices.

James Electronics
1021 Howard Avenue
San Carlos, CA 94070
Good variety of electronic components and parts. Quantity prices on some.

Kelvin Electronics Inc.
1900 New Highway
Farmingdale, NY 11735
Soldering supplies, layout materials and PC stock, etchant, wire, components, electronic kits, and variety of hardware. Quantity prices.

Kepro Circuit Systems Inc.
3630 Scarlet Oak Boulevard
St. Louis, MO 63122
Clad board and chemicals for etching. Complete line of bench-top units for PC board fabrication.

Mouser Electronics
11511 Woodside Avenue
Lakeside, CA 92040
Excellent variety of parts and components for electronics. Quantity prices.

Newark Electronics
500 N. Pulaski Road
Chicago, IL 60624
Wide variety of electronics parts, tools, equipment, and components.

Poly Paks
P.O. Box 942
South Lynnfield, MA 09140
Factory fallouts, excellent prices on grab-bag type components good for self testing. Many unusual items.

Prime Components Corp.
65 Engineers Road

Hauppauge, NY 11787
Variety of electronics parts and components.

Radio Shack
(address nearest store-check telephone directory)
Stores located nearly everywhere in USA. Generally good supply of component parts.

Techni-Tool Inc.
Apollo Road
Plymouth Meeting, PA 19462
Hand and production tools, chemicals and PC-board layout materials.

Western Components
P.O. Box 1125
Lakeside, CA 92040
Tools, test equipment, chemicals, and kits for electronics. Catalog for schools.

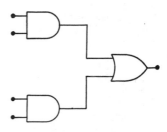

Index